Jea.
FROM THE MARKET GARDENER

Root
Vegetables

A Grower's Guide

TRANSLATED BY LAURIE BENNETT
EDITED BY PIERRE NESSMANN
ILLUSTRATIONS BY FLORE AVRAM

new society
PUBLISHERS
www.newsociety.com

© Delachaux et Niestlé, Paris, 2023. First published in France
under the title: *Les légumes racines: Les guides du jardinier maraîcher,*
Jean-Martin Fortier, Flore Avram.

The author and publisher disclaim all responsibility for any liability,
loss, or risk that may be associated with the application of any of the
contents of this book.

Inquiries regarding requests to reprint all or part of *Root Vegetables*
should be addressed to New Society Publishers at the address below.
To order directly from the publishers, please call 250-247-9737
or order online at www.newsociety.com.

Any other inquiries can be directed by mail to:
New Society Publishers P.O. Box 189, Gabriola Island, BC
V0R 1X0, Canada
(250) 247-9737

New Society Publishers is EU Compliant.
See newsociety.com for more information.

LIBRARY AND ARCHIVES CANADA CATALOGUING IN PUBLICATION
Title: Root vegetables: a grower's guide / Jean-Martin Fortier ;
    translated by Laurie Bennett;
    edited by Pierre Nessmann; illustrated by Flore Avram.
Other titles: Légumes racines. English
Names: Fortier, Jean-Martin, author | Bennett, Laurie, translator. |
    Nessmann, Pierre, editor | Avram, Flore, illustrator.
Description: Series statement: Grower's guides from the market gardener |
    Translation of: *Les légumes racines.*
Identifiers: Canadiana (print) 20250215349 | Canadiana (ebook)
    20250215357 | ISBN 9781774060179 (softcover) |
    ISBN 9781550928105 (PDF) | ISBN 9781771424066 (EPUB)
Subjects: LCSH: Root crops. | LCSH: Vegetable gardening.
Classification: LCC SB351 .F6713 2025 | DDC 635/.1—dc23

New Society Publishers' mission is to publish books that contribute
in fundamental ways to building an ecologically sustainable and
just society, and to do so with the least possible impact on the e
nvironment, in a manner that models this vision.

Creating a future where humans live in harmony with nature and with each other

Founded by Jean-Martin Fortier, the Market Gardener Institute is committed to inspiring and supporting new organic growers at every stage of their journey. Our mission is to equip them with the essential technical skills needed to thrive in their vital agricultural work.

Our vision is to multiply the number of organic, regenerative farms around the world and create a future where humans live in harmony with nature and each other.

www.themarketgardener.com

MARKET GARDENER INSTITUTE

Presenting the collection

Grower's Guides from the Market Gardener

Hi!

I am delighted to bring you this new collection of practical guides. The advice you'll find in these books is based on working methods I developed on my own microfarm and refined over the last two decades. While plenty of these concepts are not new and were passed on to me by different mentors through the years, many other ideas stem from my own farming experience. I am sure you'll come across a number of tips and tricks that are innovative, proven, and easy to implement.

Whether you are a home gardener, hobby farmer, new market gardener, or an experienced farmer looking to transition to more intensive growing on smaller plots, you will find everything you need to take your horticultural practices even further.

Wishing you success and happiness in your agricultural adventures!

Jean-Martin Fortier, market gardener in Saint-Armand, Quebec

Contents

Introduction:
A Few Words About
My Background

Drawing on principles from agroecology, permaculture, and entrepreneurship, I champion a modern form of nonmechanized farming, carried out on a human scale.

On a human scale means feeding many local families, while respecting the human and natural ecosystems in which we operate.

On a human scale means allowing market gardeners to make a decent living from their work, to run their businesses as they see fit, and to give themselves more time off than conventional farmers.

On a human scale means evolving through the use of technology but especially by relying on people and their skills and knowledge.

From Organic Farms...
I studied agroecology at McGill University's School of Environment in Montréal, where I met my wife and business partner, Maude-Hélène Desroches. At the time, we were both looking to create a new model for farming, one that would have a positive environmental impact. After graduation, we spent two years in New Mexico, USA, working on an organic farm and learning to be market gardeners.

Our microfarming aspirations were later fueled by a trip to Cuba where we spent time on *organopónicos*, fascinating urban farms that were established during the American embargo. During that era, after the fall of the USSR, the country developed a biointensive and urban agricultural model to ensure food security for the island's residents.

... to a Family-Run Microfarm

Back in Quebec in 2004, we acquired a small plot of 10 acres in Saint-Armand, in the scenic Eastern Townships. On this land, we experimented with our innovative approach to market gardening, which especially drew from the work of Eliot Coleman, an American market gardener who has been highly influential in the world of organic microfarming.

We built a 2-acre market garden, Les Jardins de la Grelinette, where we were able to test the first iterations of my method, now called the Market Gardener Method. It consists of crop rotation, the near-exclusive use of hand tools, organic growing practices, and shorter marketing channels, with direct sales made through CSA boxes and farmers' markets. At Les Jardins de la Grelinette, Maude-Hélène and I both worked full-time, and hired two farm workers (one full-time and the other part-time) to help with harvests.

Making 2 Acres Profitable

Success came quickly, both in terms of harvests and direct sales. After bringing in $33,000 in our first year, we earned twice that in the following year, and more than $110,000 in our third year of operation.

We were thus able to earn a living as market gardeners from almost the very beginning. Since then, our farm has continued to feed more than 200 families every year, offering roughly 40 types of vegetables, all grown on just 2 acres. Over the years, our harvests expanded and sales continued to increase. Eight years after starting the farm, I presented this farming model in a practical guide called *The Market Gardener* in 2014. The book was an instant success — over 250,000 copies have now been sold, and it has been translated into nine languages.

In 2015, with the support of a generous patron, I founded Ferme des Quatre-Temps in Hemmingford, Quebec, with the vision of creating a model for the future of ecological agriculture. On this 160-acre farm, we established a polyculture system in a closed-loop cycle, raising pasture-fed cattle, pigs, and hens, alongside a culinary laboratory. At the heart of the farm, 7.5 acres were dedicated to a market garden, where we applied the growing methods developed at Les Jardins de la Grelinette. It is here that I teach my apprentices the principles of productive and profitable market gardening.

The project was featured in a TV show called *Les fermiers*, which follows the evolution of Ferme des Quatre-Temps and its apprentices, who later start their own farms in front of the cameras. The show was a hit in Quebec and is now available on TV5 Monde and Apple TV.

In parallel, I worked to expand my methods to reach a broader, global audience. In 2018, we launched the Market Gardener Masterclass, a fully online course now available in over 90 countries. To further support this initiative, I founded the Market Gardener Institute with a clear mission: to educate the next generation of growers by equipping them with the knowledge, skills, and resources needed to become leaders in the organic farming movement.

The Institute has two key objectives: to teach best practices in market gardening techniques and growing methods, and to demonstrate that small-scale farming worldwide can not only be ecological but also productive and profitable. On a global scale, it's the number of farms, not their size, that holds the key to feeding the world.

Inspiring Change

My ambition is to drive meaningful change in society by promoting a way of farming that honors nature, supports communities, and empowers local farmers. I believe in a decentralized farming model, built farm by farm, as the foundation for a truly sustainable and resilient food system.

Since 2020, I have proudly served as an ambassador for the prestigious Rodale Institute, which researches regenerative organic farming practices in the United States and beyond. I am also honored to be the ambassador for Growers and Co., a company that develops tools and apparel for new organic growers. In 2023, we launched Espace Old Mill, a restaurant and market garden set in one location. The restaurant uses the best produce in the region, including harvests from our own farm.

What Is the
Market Gardener Method?

While my approach may seem innovative, it is founded on practices that were first developed by 19th-century Parisian gardeners, who fed more than two million people through a network of thousands of market gardens—precursors to our modern-day microfarms—within the city of Paris.

These market gardeners applied remarkable ingenuity, skills, and knowledge to meet the increasing food demands of a city in the midst of urbanization and demographic expansion. They achieved this through organic, nonmechanized agriculture. From the mid-18th century to the 20th century, many books were written about the innovative practices of these market gardeners, whose technical feats were admired throughout Europe. But with the advent of modern practices, much of this know-how was relegated to the past.

As a result of mechanization, the advent of agronomic science, and improved refrigeration and transport that brought in fresh and inexpensive food grown abroad, farms grew in size, became less diversified, and took on a more technological focus—a trend that continues today.

Fortunately, these inspiring models led to the development of horticultural methods that have endured, and with the same objective: to grow sustainably, by maximizing vegetable yields without degrading soil quality. We now use the term "biointensive" to describe these methods. Unlike extensive agricultural operations, they continue to work on a human scale and offer farmers the opportunity to use little mechanization. Despite what some may believe, this approach is also profitable.

By working on only small plots of land, market gardeners can keep start-up investments to a minimum, compared to the funds needed for a conventional farm. Biointensive farmers also require a smaller workforce, doing the work themselves with the help of just a few employees. They also sell their produce directly to customers, avoiding commissions to intermediaries. These three factors allow market gardeners to start generating profits quickly.

Still, it's important to remember that working the land is never easy. While market gardeners can make a good living with this method, the first seasons are time-consuming and require a significant workload and financial investment. In this profession, nothing comes easy, and every dollar you earn is the fruit of your labor, the result of your organizational skills. That's why I always tell my apprentices to learn how to work smarter, not harder.

From a financial perspective, market gardeners should plan to start with an investment of $50,000 to $150,000, depending on whether certain assets are already available—such as a building that can be converted, access to abundant water, electricity, natural gas, or a vehicle. This amount does not include the cost of purchasing land, which can be amortized over 20 years, if needed. Renting is also an option that can prove very profitable, especially when the farm is located near a city or an affluent municipality, where land is expensive.

Regardless of experience and preparation, the first years of market gardening will be intense. Opening new ground, constructing greenhouses and tunnels, and setting up infrastructure (irrigation, washing and packing stations, nurseries, etc.) all take extra time and effort. However, once this phase is complete, market gardeners who have mastered their craft can do more than just make a living off a few acres—they can earn a very decent living.

This leads to another key principle I teach: your farm should work for you, not the other way around. Profitable and productive farming is possible, but you need to set it up for success.

Preface:
Unearthing a World of
Root Vegetables

I am thrilled to present the third installment of the Grower's Guides from the Market Gardener, which is all about the world of root vegetables. Personally, they are my favorites. Root vegetables are true gems growing in the depths of our fertile soils. They offer a wide array of flavors and textures just waiting to be explored. Some, like carrots, beets, and potatoes, are familiar classics, while others, like kohlrabi, winter radishes, and sweet potatoes, deserve more attention. Of course, we can't forget oca and root parsley, little-known treasures that would undoubtedly surprise and impress your friends and family!

Whether you are a keen home gardener looking to improve your yields in a small space or an experienced market gardener seeking out new opportunities, root vegetables are a sound choice. Because these vegetables grow underground, they allow growers to make the most of their available space while delivering bountiful and nutritious harvests. This practical guide will show you the secrets to growing root crops, from careful soil preparation to harvest.

You will learn precise seeding and transplanting techniques, as well as best practices to care for your crops and keep them healthy. To finish, I will share some storage tips, so that you can enjoy these delightful vegetables all year-round.

Now, let these extraordinary vegetables inspire you, and remember that the soil rewards skillful hands.

Happy gardening!

Jean-Martin Fortier, market gardener in Saint-Armand, Quebec

Root Vegetables: The Essentials

Vegetables that grow underground—grouped together under the sizable umbrella term "root vegetables," even though some are not in fact roots—include crops that are essential to our diet and other lesser-known plants with subtle flavors that make them just as precious. They are all treasures of the earth.

Vegetables tend to be classified by botanists and market gardeners according to the part of the plant that they represent, which is generally the part that we eat. For example, we eat the leaves of leafy green vegetables like lettuce and spinach, whereas with fruit vegetables like tomatoes or melons, we consume the fruit. When it comes to root vegetables (which, mind you, are not always true roots!), we eat the parts that grow underground.

All plants have roots, and the root system is essential for plant growth. It acts as an anchor in the soil, allowing the plant to withstand wind and to bear branches and foliage, as well as flowers and fruits. This dense network of roots (each one a different size) can make its way into even the smallest crevices in the soil, allowing the plant to absorb the nutrients (soil minerals) and water needed to develop aboveground organs like stems, leaves, flowers, and fruits.

Not all underground plant organs are edible. Some are unfit for consumption and may even be toxic, so it's best to get more information before eating them. Others have been identified as edible since the dawn of time, and some of them are especially interesting because they are flavorful, easy to grow, or quite simply, high-yielding. The underground organs of plants known as root vegetables are among these classics.

Botanically speaking, not all underground organs are roots. The catch-all term "root vegetables" refers to all vegetable species with organs that grow underground and at the soil surface. However, the parts of the plant that we eat are not always roots in the true sense of the word. So, we need to distinguish true roots from plant organs that also grow underground or at the soil surface that aren't actually true roots.

The two vegetables described in this book that are most similar to a true root, i.e., a root that is not a storage organ, are scorzonera (a black root) and salsify (a brownish off-white root sometimes called vegetable oyster). These slender taproots are protected by thick black or brownish skin, with an appearance and flavor that have changed very little over the centuries.

Another category that is quite similar, from a botanical perspective, includes carrots, parsnips, turnip-rooted chervil, and root parsley, which are also known as conical roots or storage roots. These root vegetables can be distinguished from scorzonera and salsify by their slightly thicker and fleshier taproots, which serve as storage organs and can be found underground or partially buried. In the plant world, the process by which taproots become enlarged is referred to as root thickening or storage root formation. This allows the plants to build up nutrient reserves, especially to prepare for hard times ahead, like a cold season in our region. As soon as milder weather returns, they can regrow by drawing upon stored nutrients.

Carrots   Parsnips   Parsley tubers

These reserves lend the roots a thicker and fleshier appearance, and when the skin is colored, they look more appetizing. Beyond their physical appearance, storage roots also have an appealing flavor as the reserves are primarily made up of tasty sugars.

The word "tuber" comes from the Latin *tuberculum*, meaning "small swelling." In root vegetables, this refers to underground organs that appear enlarged, range in size, and store nutrients with a high starch content. Tubers grow along the roots, which keep them all connected. Unlike the taproots of true roots and storage roots, when dormant, tubers have all the necessary organs to develop a new plant if separated from the mother plant. Thus, each tuber is capable of producing roots, stems, leaves, and flowers that would allow it to be self-sufficient and, most importantly, to develop a plant identical to the one it originated from. Oca, mashua, sweet potatoes, potatoes, and yacon produce tubers with good flavor, high yields, and a long shelf life, which have made them some of the world's leading root vegetables.

Rhizomes, often found in flowering perennials like irises or shrubs like bamboo, are underground organs that grow horizontally. *Helianthus strumosus* (paleleaf woodland sunflower) and Jerusalem artichokes are two that have an edible rhizome, so they are considered to be vegetables. In fact, they are quite often listed as both perennial vegetables and perennial decorative plants, grown for their flowers or foliage. Botanically speaking, rhizomes are underground stems. They are cylindrical, range in thickness, and have a dented surface that features rough dark skin, in general, quite similar to roots (for which they may be mistaken). However, rhizomes can be distinguished from roots by the fact that they have buds and small roots that allow them to grow new stems, leaves, and flowers every year. Such plants are therefore perennial and independent organs that, if cut into fragments, may continue growing and produce a new plant. This can make them a little invasive, especially if they are not killed off by frost.

The last category of root vegetables is not in fact a root since the edible organ, the hypocotyl, grows at the soil surface, well above the root system. It sits between the base of a plant, called the crown (where the root and shoot systems meet), and the cotyledons (the first 2 leaves to appear during germination). This category has the largest number of root vegetables: beets, kohlrabi, celeriac, turnips, rutabagas, and radishes, including daikon radishes. The flesh of these root vegetables, although protected by a layer of skin, is more exposed to light as the storage organ grows aboveground. These crops should be covered with soil through regular hilling to prevent cracking and hardening.

Beet   Kohlrabi   Radish

Note: Bulb-producing vegetables like garlic, onions, and shallots are a separate category from root vegetables because the underground organ consists of leaves compressed into densely packed scales around a vegetative growing point. Edible bulbs used in vegetable gardens will be the subject of a specific book in this collection.

Root vegetables are mainly intended for human consumption, but some species are cultivated specifically for livestock. These are known as fodder roots or forage roots.

You might, understandably, wonder why we eat these underground organs that don't look particularly appetizing at first glance. Some of them, like turnip-rooted chervil, root parsley, Jerusalem artichokes, and salsify, could have remained a marginal part of our diets, or even been forgotten—and they were, for a while—but they are making a comeback, showcased and celebrated by creative chefs. So how did root vegetables come to be so successful? It is likely due to their origins.

Humans have probably been eating underground plant organs since the Neanderthal period. When analyzing the teeth of Neanderthals, researchers found both plant matter and tuber residue. Later, in European prehistory down through the Middle Ages, the harvest and consumption of edible roots from wild plants—initially gathered in nature and later domesticated and cultivated—were common practices. Don't forget that the variety of vegetables we know today did not exist then. Most fruiting vegetables, such as tomatoes or eggplants, which were brought to Europe from South America in the 17th century, were not yet a part of daily diets. People simply gathered and ate leaves, wild fruits, and roots. Since roots were relatively fibrous, offered little sustenance, and had poor flavor, we can speculate that they were primarily used to provide a little variety and perhaps some beneficial nutrients.

Root vegetable cultivation continued and developed further most likely due to improvements made to these crops over the centuries. Unlike leafy vegetables, characterized by their freshness and appetizing colors, most root crops probably looked unenticing (filiform organs, encased in rough, cracked, and dark skin—you really had to be hungry to eat them!). But they kept well and could therefore be consumed long after harvesting, when they were far from fresh. And thus, they became the staple of many diets, for centuries, and especially in times of famine and war, as they were easy and inexpensive to cultivate.

It was only much later, starting in the 17th century, and then in the 18th and 19th centuries, that the roots increased in color, flavor, and yields, thanks to crossbreeding with species that came from other continents in the luggage of traveling botanists. Carrots are the best example of this change. Originally, wild carrots were white, purple, and red, but not orange. Orange varieties first appeared in the 17th century when they were created by Dutch seed producers looking to pay tribute to the House of Orange-Nassau, a family that ruled the United Provinces, known today as the Netherlands. To achieve this, they crossed red wild species with yellow wild species from the East. Over the centuries, improvements in root vegetables multiplied, and even today, seed farmers are constantly improving and perfecting these underground plant organs, elevating them to the status of treasures of the earth.

Root vegetables tend to grow slowly and develop inconspicuously, hidden in the bowels of the earth. And once harvested, their sometimes unattractive appearance is not the most appetizing. And yet root vegetables have been dietary staples since the dawn of time. The most common ones—with potatoes in the lead—are cooked year-round. Others, such as radishes, beets, celeriac, and turnips, make seasonal appearances at market stalls. Some, rarer or forgotten, like salsify and parsnips, are rich with subtle flavors that are now being reimagined and showcased by the greatest Michelin Star chefs.

Twenty root vegetables in particular are high on my list of must-have crops. They deserve to be allotted 1 or 2 beds—or even more—on a vegetable farm, or a few rows in home gardens. In the pages that follow, you will find 14 classic root vegetables that we grow and sell at markets and are used in our restaurant. Each one comes with a detailed description of growing, harvesting, and storage methods. You will also find 6 more unusual crops, which I encourage you to try.

Jean-Martin Fortier's 20 Favorite Root Vegetables

Data Sheet

COMMON NAMES: Beet, table beet, red beet, beetroot, garden beet.

SCIENTIFIC NAME: *Beta vulgaris* subsp. *vulgaris*.

FAMILY: Amaranthaceae.

REQUIREMENTS: Undemanding, adapts well to changes in temperature, easy to grow. Sensitive to boron deficiencies, which causes black spots in beets. Prefers loose, deep, organically rich, and fast-warming soil.

SPACING: 3 rows per bed, set 8 inches (20 cm) apart with a 5-inch (12 cm) in-row spacing (after thinning).

SEEDING: From mid-March to late June, early July.

DAYS TO MATURITY: 50 to 60 days (early and main-season beets), 100 to 120 days (storage beets).

ENEMIES: Cercosporiosis (foliar disease), beet leaf miner, slugs in early crops, and rodents (pre-harvest).

VARIETIES: Boro, Red Ace, Moneta (monogerm), Early Wonder, Touchstone Gold (yellow), Chioggia (our favorite).

NOTE FROM JEAN-MARTIN FORTIER
Traditionally, beets have been known as winter root vegetables, valued for their sweet taste, which is especially comforting in the height of the cold season. However, they are increasingly eaten in spring and summer, especially grated raw or for the young leaves.

Beets

Sea beets *(Beta vulgaris* var. *maritima)*, were native to Southern Europe and cultivated by Egyptians and Romans, but the varieties we know today, selected for their flavor or root color, appeared much later, in the 19th century. Beets are biennial species, with a growth cycle that typically occurs over 2 years. However, because the edible part of the plant develops in the first year, it is grown in a single year, like an annual.

While beets grown for fodder or sugar develop larger roots, those grown as root vegetables (garden beets) are more rounded, sometimes flat, or cylindrical. They grow at the soil surface and produce a rosette of green or red stalks topped by green and sometimes red leaves with very pronounced and colorful veins. The flesh is red, but varieties range from pale yellow to an orangish-yellow. This root vegetable contains pigments, betanin and anthocyanin, that are used as dyes in the food industry. Garden beets can be eaten raw and grated, or steamed, cooled, and cubed or sliced, then tossed with a vinaigrette. A cross-section of the root reveals concentric circles, which are characteristic of this vegetable. Young, still-tender leaves can be eaten in salads, like mesclun.

Beets

⏤⏤⏤⏤⏤⏤⏤⏤⏤◦⏤⏤⏤⏤⏤⏤⏤⏤⏤

**"Although beets can be a little fussy, they are actually easy
to grow. Eaten raw or cooked, this root vegetable is slowly
making its way into our kitchens!"**
Jean-Martin Fortier

Planting

Preparing the Soil

Loosen the soil with the broadfork, then
apply a fertilizer mixture made of compost,
alfalfa flour (2-0-2), and dehydrated chicken
manure (5-3-2). Use a power harrow set to a
depth of 1.5 inches (4 cm) and a bed
preparation rake to remove debris (rocks,
clods of soil, crop residues) that might
interfere with seeding.

Seeding

For early and main-season crops, direct
seed beets in a single row or in 3 rows,
from mid-March (with a row cover to
protect against late frosts) until mid-May.
Beets intended for storage are sown in June
and July so that they will reach maturity
roughly 1 week before the first fall frost
dates. Beet seeds will be in the ground for
several weeks, so they need clean, weed-free
soil. Use the stale seedbed technique ahead
of time to prepare the beds.

To avoid thinning, which is time-
consuming, it is preferable to seed with a
single-row seeder (Jang JP-1). Start by
seeding the middle row, then seed the outer
rows 8 inches (20 cm) away from it.

Caring for Seedlings

Water immediately after seeding, then keep the soil moist for the next 7 to 10 days until the seeds germinate. Afterwards, be very consistent with watering because a lack of water in hot weather can cause the crop to go to seed and lead to root cracking, in addition to making the flesh more fibrous.

Beet Seeds

These seeds do not have a typical smooth and even shape. They feature several angular facets and are called clusters, each containing 3 to 7 germs (multigerm). If the clusters have not been split apart, through genetic or mechanical intervention, the variety is referred to as multigerm (2 to 4 germs), double-germ (1 to 2 germs), or monogerm (1 germ). Germination occurs within 8 to 10 days when temperatures are 64°F to 68°F (18–20 °C).

Tip from Jean-Martin Fortier

Beets are particularly sensitive to boron deficiency that affects crop development and causes black spots to form in the flesh.

Apply boron twice as a precaution: once 3 weeks after seeding, then again when the beets are 1 inch (2.5 cm) tall. These treatments can be combined with an algae solution application. It's best to apply the boron early in the morning to encourage absorption of the solution by the leaves.

Maintenance

Weeding

Run a flex tine weeder down the bed as soon as
the plants have grown a true leaf, about 10 to 15
days after germination, to loosen the soil while
weeding both in-row and between rows in a
single pass.

Weed again 10 to 15 days later using a wheel
hoe with bio-discs so that you can weed
both along the row and in the row in a
single pass. Then, 7 to 10 days after using
the bio-discs, and before weeds form 2
leaves, cultivate the bed with a 5-inch
(13.5 cm) stirrup hoe.

You can also take this opportunity to thin
the beets, with the hoe or by hand
(opposite), leaving about 5 inches (12 cm)
between them.

Plant Protection

Cercospora leaf spot is a common foliar disease in the Amaranthaceae family. It emerges when temperatures are hot and humidity runs high. For disease prevention, establish a crop rotation and wait 2 to 3 years before growing beets again in any given bed. It should be noted that using drip irrigation avoids wetting the foliage and creates a drier environment that is less favorable to this disease.

Usually, the beet leaf miner is no longer a problem by the end of summer as growers cut the foliage off storage beets. A certain level of purely aesthetic damage can thus be tolerated as long as it does not affect yields. However, with spring crops—especially those grown for baby greens used in mesclun—it's a good idea to cover beds with insect netting.

Harvest

Early and Main-Season Beets

Beets can be cultivated practically year-round, allowing growers to harvest young leaves to be eaten raw in salads or cooked like spinach. When harvesting early and main-season beets to be eaten raw, lift them with a spading fork and pull the plants out by gently grasping the stems. They are small, only about 2 inches (5–6 cm) in diameter, tender, sweet, and remarkably tasty. Gather them into bunches for market or store them loose at home.

Storage Beets

Harvest storage beets when the weather is dry. Use a broadfork to lift all the roots at the same time, then remove the foliage and leave it on the bed. Place the beets in a harvest crate.

At this stage, the sorting and quality control process must be rigorous. Remove any beets showing signs of disease or injury, as well as those smaller than 2 inches (5 cm) in diameter, and store them separately, to be sold or eaten sooner.

Storage

Do not clean these beets during harvest, as they should be stored unwashed. Home gardeners keep them in sand, while professionals use a cold room maintained between 36°F and 39°F (2–4°C). Later, take the roots out of storage and soak them in water for a few minutes to remove the dirt.

Tip from Jean-Martin Fortier

Depending on how long you are storing beets, you may need to sort through them and remove damaged roots so that they will not affect the rest of the harvest.

Data Sheet

COMMON NAME: Carrot. We distinguish between early baby carrots and later storage carrots.

SCIENTIFIC NAME: *Daucus carota* subsp. *sativus*.

FAMILY: Apiaceae.

REQUIREMENTS: Fairly easy to cultivate, carrots grow best in loose, deep, sandy soil that promotes development of uniform roots. Fresh manure can be harmful, as well as overly heavy soils that become waterlogged in the spring and fall.

SPACING: 6 rows per bed set 5 inches (12.5 cm) apart, with an in-row spacing of about 2.5 inches (6 cm).

SEEDING: From mid-March to late June, early July.

DAYS TO MATURITY: 60 to 70 days (baby carrots), 80 to 90 days (storage carrots).

ENEMIES: Carrot fly *(Psila rosae)* and early blight or Alternaria leaf blight *(Alternaria dauci)*.

VARIETIES: Bolero, Yaya, and Nelson (Nantes-style storage carrot), Napoli (our favorite for fall), and the colorful varieties Purple Haze (purple), Romance (red), Creampak (light yellow), Rubyprince (carmine pink), and Snow Man (white).

NOTE FROM JEAN-MARTIN FORTIER Carrots are the vegetables that best highlight the difference between produce grown on local, organic regenerative farms and produce from industrial agriculture. The contrast in flavor is so striking that carrots are the perfect crop to convince people of the benefits of organic farming.

Carrots

While the carrot is a biennial species that grows wild in France, the root we cultivate today is native to Iran, where it was improved over the years to feature colors ranging from yellow and red to a near-black purple. In Europe, it underwent a selective breeding process in the 17th century, which crossed white wild species and colored varieties, while orange garden varieties are the result of Dutch breeding in the 18th and 19th centuries.

This vegetable produces delicate, fragrant foliage (tops), supported by stems that extend from a taproot, which can be round, tapered, or cylindrical, ranging in size: short, medium, and long. It consists of a tender core, usually lighter in color than the outer skin. In Europe, many regions, such as Brittany, Pays de la Loire, Alsace, and Manche, have light soils, which make them ideal for carrot cultivation. This is why some varieties were named after these places. Carrots are rich in antioxidants, have strong anti-inflammatory properties, and are known to be a good source of fiber, minerals, provitamins, and vitamins A, C, and B. Their orange hue is due to the presence of carotene—an orange pigment—whereas purple varieties contain anthocyanins. All carrots can be consumed cooked, but eating them raw is the best way to benefit from their nutritional properties.

Baby Carrots

—————●—————

**"I love early baby carrots because
they are tender and sweet!"**
Jean-Martin Fortier

Planting

Preparing the Soil

Baby carrots grow best in soil that warms
up quickly and, most importantly, has been
properly loosened. It should ideally be
sandy, so that young roots can develop
unhindered, with uniform sizing.
After preparing the bed with a broadfork
and a power harrow set to a depth of
roughly 1 inch (2–3 cm), level the soil
surface with a bed preparation rake for a
flat and even seedbed.

Seeding

Stagger carrot seedings from late February
to mid-April, under a caterpillar tunnel or
low tunnel, by sowing directly into the
ground. Sow densely, at 5 or 6 rows per bed.
To achieve this spacing, use a Jang seeder or,
for smaller quantities, a dial seed sower,
especially in home gardens. First stake a
string down the bed and make it taut. Then
use a tool handle to dig a furrow about
0.5 inches (1 cm) deep for the seeds. Starting
in early April, if the soil is warm enough,
direct sow carrots under a floating row
cover to help with germination.

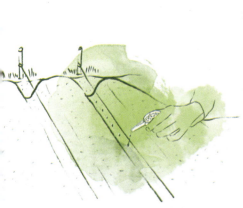

Thinning

If working with a dial seed sower, you will need to thin the crop, removing plants that are too close together and may cause narrow or misshapen roots. Once the seedlings have 3 or 4 well-formed leaves, pull out the frailest, leaving one plant every 1.5 to 2 inches (4–5 cm).

Weeding

Weed control is the biggest challenge when growing baby carrots. Seeds take between 8 and 20 days to germinate, which gives weeds time to get established. We therefore recommend using the stale seedbed technique in which growers allow weed seeds to germinate and then remove them before sowing carrots. We also recommend using a flex tine weeder on the bed about 10 days after sowing. Ten to 15 days later, when the tops are 3 to 4 inches tall (8–10 cm), hoe carrots with bio-discs, running down each row once to weed both the bed and the aisle. Remove any remaining weeds by hand.

Maintenance

Plant Protection

When it comes to pests and diseases, growing carrots isn't without its challenges! Two types of pests—carrot flies and carrot weevils—bore into the roots and cause scarring. The most effective solution is to use insect netting starting in April when adult insect flight begins. Lay the netting over metal hoops to avoid hindering foliage growth, and make sure the sides are held down properly for a perfect seal.

Tip from Jean-Martin Fortier

For evenly shaped roots, the crop must be watered regularly and the ground must be moist and cool, even at depth. This is why we recommend setting up a mini-sprinkler system to water the beds. Of course, pay attention to irrigation frequency: in rainy weather, you must hold off on watering because too much water, just like an irregular supply, can cause cracking or splitting.

Harvest

Baby carrots are harvested young, when the roots are only about 6 inches (15 cm) long, but they can even be harvested at 3 to 4 inches (8–10 cm).

In a first pass down the row, harvest the roots that are most mature and the best shape. This selection process leaves more room for smaller plants to grow and be harvested later.

In subsequent harvests, carrots that have spent more time in the ground are often longer (at least 6 inches, 15 cm) and harder to remove without damaging them. You should therefore wait until the soil has been loosened by irrigation or rain before harvesting. When you grasp the top and gently rotate the root, it should easily come out of the ground. If you encounter any resistance, it's best to loosen the soil first with a broadfork or a spading fork. This will make it easier to extract the carrots.

Storage

It is better to eat baby carrots as soon as they are harvested because this is when they are most tasty, tender, and fragrant. Avoid storing them for too long. They will keep for a few days in a cool space (between 40°F and 43°F, 4–6°C), but make sure to remove the foliage, which tends to dehydrate the roots and make them lose that remarkably tender texture.

Storage Carrots

"Do as I do, and harvest storage carrots after the first frosts because a little cold exposure will make them sweeter!"
Jean-Martin Fortier

Planting

Preparing the Soil

When you need to sow carrots right after harvesting another crop, lay an occultation tarp over the bed. Two to 3 weeks later, remove the tarp and rake off the largest pieces of debris, then loosen the bed with a broadfork. Next, use a rotary harrow to cultivate and level the soil to a depth of about 1 inch (2–3 cm), then carefully smooth the surface with a bed preparation rake.

Seeding Into a Bed of Compost

1 Spread roughly 1 inch (2–3 cm) of weed-free compost over the width of the bed without incorporating it into the soil (otherwise, weeds may germinate quickly).

2 Fill and calibrate your Jang seeder and begin sowing the crop, roughly 1 inch (3 cm) apart. After the first pass, compare the weight of the seeds you planted with the estimated weight to be sown, then adjust seeder settings as needed.

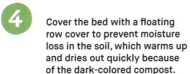

3 After seeding the crop, set up a mini-sprinkler system and make sure to keep the soil moist for the next 10 days to promote good germination.

4 Cover the bed with a floating row cover to prevent moisture loss in the soil, which warms up and dries out quickly because of the dark-colored compost.

Tip from Jean-Martin Fortier

Storage carrots differ from baby carrots in that they spend more time in the ground, and once stored, they can be eaten practically all year-round. For your crop rotation plan, we recommend seeding storage carrots from mid-May to July in beds left empty, for example, after garlic harvest so they will benefit from the mature compost that was applied the previous fall before planting garlic.

Caring for Seedlings

Even if you seed onto compost or flame weed the bed before germination, it is essential that you monitor the crop throughout the emergence phase. As with baby carrots, we recommend running a flex tine weeder down the bed 10 days after sowing and using a hoe with bio-discs 10 to 15 days later, when the leaves are 3 to 4 inches (8–10 cm) long. With the bio-discs, a single pass per row allows you to weed both the bed and the aisle. If any large weeds remain after hoeing, remove the biggest ones by hand.

Tip from Jean-Martin Fortier

Weeds that germinate after the last bio-disc pass will compete less with young carrots that begin to fill out the bed. Even if weeds manage to grow, they will not have time to go to seed before the carrot harvest ends. This is why we typically do not plan for any subsequent weeding. However, if some weeds do go to seed, we recommend pulling them out by hand.

Maintenance

The steps to take during cultivation are identical for storage carrots and baby carrots. Irrigate regularly and make sure the roots are not subjected to intermittent watering, otherwise, they will crack. After several waterings or heavy rains that cause surface compaction, you must loosen the soil.

With the flex tine weeder, you can cultivate the soil surface without damaging crop roots and foliage. Lastly, as with baby carrots, insect netting is recommended to effectively control carrot flies as the nuptial flights of adult insects occur from May through to the end of October.

Harvest

Harvesting storage carrots is labor-intensive because, depending on the quantities seeded, the number of carrots to lift may be significant. This operation should be carried out all at once, usually after one or two frosts that increase their sugar content. Harvest when the soil is loose and cool but not wet and sticky. Use a broadfork to loosen the bed and gently lift the soil along each row, then pull the carrots and shake off the dirt, before laying them on the ground side by side.

Storage

When harvest crates come back from the field, brush soil off the carrots and then take them to a cold room kept at 36°F (2°C). Store the roots as they are, unwashed, to extend their shelf life to around 6 months.

They should only be washed right before being sold or eaten. Start by quickly hosing down the carrots, then clean them more thoroughly with a root washer.

Next, sort the harvest, keeping only those that are in very good condition, then top and place them in harvest crates. Set aside damaged roots, so they can quickly be eaten, juiced, or canned.

Tip from Jean-Martin Fortier

In mild climates, carrots can be stored in the ground because they can withstand temperatures as low as 21°F (-6°C). As a precaution, you can cover the soil with straw or dead leaves to mitigate the effects of frost and make it easier to drive a broadfork into the ground to lift the crop.

Home gardeners can also store carrots in a root clamp. Put the carrots in crates filled with wet sand to prevent the roots from drying out. Then store them in a dark frost-free cellar, where they will keep for 3 to 4 months.

Data Sheet

COMMON NAMES: Celeriac, celery root, knob celery, turnip-rooted celery.
SCIENTIFIC NAME: *Apium graveolens* var. *rapaceum.*
FAMILY: *Apiaceae.*

REQUIREMENTS: Celeriac grows best in cool climates and does not tolerate drought and inconsistent watering. It thrives in loose, deep, and very rich soil as it requires a lot of organic matter to develop a nice enlarged root.

SPACING: 3 rows per bed set 8 inches (20 cm) apart with a 12-inch (30 cm) in-row spacing.
SEEDING: From February to mid-March.
PLANTING: In May.
DAYS TO MATURITY: 190 to 200 days.
ENEMIES: Septoria leaf spot (black spots) and rust can develop on the leaves, while the root can be susceptible to various types of rot. Carrot and celery flies, as well as aphids—which spread viruses (mosaic)—and cutworms, will attack this crop but without seriously impacting yields.

VARIETIES: Mars (Septoria leaf spot resistant and our favorite), Brilliant, Monarch.
NOTE FROM JEAN-MARTIN FORTIER
Be patient! This is a slow-growing crop, but it has excellent storage qualities and will keep for a long time, making it a must-have root vegetable and a great complement to other winter produce.

Celeriac (Celery Root)

Celery is a biennial species originating in the Mediterranean where it was known to both the Greeks and Egyptians. In Northern Europe, it was introduced and cultivated as early as the Middle Ages, but it was in the 17th century that the crop truly took off in two forms: celeriac and stalk celery. Here, we are focusing on celeriac, also known as celery root. The part we eat is not actually a root but is instead a tuberous stem (an enlargement of the hypocotyl) that is often mistaken for a root because of its color and appearance. The celeriac hypocotyl is round and fleshy, topped by ribbed stalks and green leaves that can grow 16 to 20 inches (40–50 cm) tall. The shoot system is quite flavorful and can be eaten. Celeriac is a fall and winter vegetable that grows slowly and monopolizes garden space. It freezes at about 25°F (-4°C) and must be harvested once the first frosts arrive and stored in a cellar.

Celeriac is high in vitamin E and minerals, and has stimulating properties due to the volatile oil stored in the plant tissue. The flesh is firm, white, and quickly oxidizes when exposed to air, so it is best suited to salads and rémoulades. It can also be baked, fried, or steamed, eaten in soups or gratins, or added to a mixed vegetable dish.

Celeriac

———•———

"While celeriac is easier to grow than its cousin, celery, it remains quite challenging and requires a fair amount of attention, especially in the first few weeks."

Jean-Martin Fortier

Planting

Seeding Under Shelter

Because celeriac is quite slow to germinate, it must be sown under cover. For professional growers, this means in a greenhouse, while home gardeners can use a mini greenhouse. Exposure to light is essential for even germination. Place seeded flats on heating mats to keep the soil temperature at 75°F (24°C) until the seeds come up. As soon as shoots emerge, 2 to 3 weeks after seeding, reduce overnight temperatures to 64°F (18°C). Keep the soil moist at all times.

Potting Up

About 4 to 5 weeks after sowing celeriac, when the seedlings have 2 pairs of true leaves, pot them up. One by one, transfer plants to plug flats or pots where they will grow for a few more weeks. You can generally expect 70 days from seeding to transplant.

Tip from Jean-Martin Fortier

About 7 to 10 days before transplanting seedlings into the field, it's a good idea to harden them off and to reduce watering. However, do not drop the temperature too much, as seedlings kept below 55°F (13°C) for even 10 days may bolt.

Preparing the Soil

Use a broadfork to loosen the soil, then apply compost. Run a power harrow down the bed at a depth of 2 inches (5 cm). To make it easier to transplant seedlings, prepare the soil right before planting. Seedlings should be ready when the foliage is 4 to 6 inches tall (10–15 cm), roughly 8 to 10 weeks after the crop is sown.

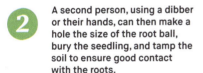

Planting

1 Before planting, use a row marker roller to indicate where each seedling will go. One person can drop the plants where the lines intersect.

2 A second person, using a dibber or their hands, can then make a hole the size of the root ball, bury the seedling, and tamp the soil to ensure good contact with the roots.

3 Water the base of each plant to help the soil settle around the root ball, then set up a drip irrigation system with one line of tape per row. Make sure the soil is moist at a good depth.

Maintenance

Weeding

About 2 weeks after transplanting the crop, run a wheel hoe with 2 pairs of bio-discs down the bed to weed and hill the celeriac row in a single pass. Repeat this operation every 1 or 2 weeks until the roots have become so big that the discs may damage them. The advantage of this tool, which features 2 pairs of discs, is that it allows growers to hill the plants. This in turn limits the roots' sun exposure and encourages blanching.

Plant Protection

Celeriac is susceptible to celery mosaic virus that causes leaf curl and yellow mottling on foliage. This virus is spread by aphids, and regular weeding will keep them away from the crop. As a preventive measure, septoria leaf spot can be limited by applying calcium dissolved in irrigation water throughout the growth cycle.

Tip from Jean-Martin Fortier

When the use of bio-discs is no longer possible, we opt for the collinear hoe to weed in between rows. If any weeds remain, pull them out by hand.

Harvest

Lifting the Crop

Harvest celeriac from late summer to fall once the root is 3 to 6 inches (8–15 cm) in diameter. Loosen the soil by driving a spading fork into the ground between the rows, lifting both soil and roots. Grasp the foliage to pull out the celeriac, then cut the roots flush with the bulb.

Remove the leaves, cutting the stalks down to about ½ inch (1 cm). To finish, place the roots in a harvest crate to prepare for storage.

Storage

When celeriac comes in from the field, soak it in cold water to dissolve and loosen any soil that is still stuck to the skin, then hose it down with a powerful water jet.

Let it air-dry, then store the harvest in crates or bins in a cold room maintained at 36°F (2°C), where it will keep for 4 to 6 months.

Tip from Jean-Martin Fortier

Home gardeners working in a mild climate can store celeriac outdoors in a root clamp or in a pit. In colder climates, it is best to store them in bins in a frost-free, dark, cool, ventilated space. With both methods, lay the roots side by side on a bed of sand, covering each layer with leaves or straw.

Data Sheet

COMMON NAMES: Kohlrabi, turnip cabbage, stem turnip, German turnip.

SCIENTIFIC NAME: *Brassica oleracea* L. var. *gongylodes.*

FAMILY: Brassicaceae.

REQUIREMENTS: Kohlrabi is fairly undemanding and grows in good garden soil that is loose and fertile, but not overly rich. It does not tolerate hot and dry spells, which make the flesh less tender and more pungent.

SPACING: 3 rows per bed set 8 to 10 inches (20–25 cm) apart, with an in-row spacing of about 8 inches (20 cm).

4 rows per bed for intensive cultivation.

SEEDING: Staggered from January to February for a spring harvest, then July to August for a fall and winter harvest.

PLANTING: Mid-February to April (spring crop) and September (fall crop).

DAYS TO MATURITY: 40 days after transplanting in the field, 75 to 80 days after seeding.

ENEMIES: Flea beetle, gall midge, and cabbage moth harm both spring and fall crops.

VARIETIES: Korridor, Kolibri, and Azur Star (both deep purple), Korist (ideal for spring), Superschmelz and Kossak (our favorite varieties for storage).

NOTE FROM JEAN-MARTIN FORTIER
While kohlrabi flesh will become quite woody and look glassy after a hard frost, it is remarkably hardy in low temperatures. After a period of frost, harvests are tastier and more tender.

Kohlrabi

Kohlrabi, sometimes called German turnip and turnip cabbage, is a biennial species typically grown as an annual. Native to Central Europe and the Mediterranean region, it has been known since antiquity and was cited among vegetable species in the *Capitulare de villis*, an administrative text dating from the 8th or 9th century, possibly issued during the reign of Charlemagne. In the 19th century, the popularity of kohlrabi soared when it made the jump from vegetable gardens to large-scale farming.

Kohlrabi holds a special place in the cabbage family because of its swollen stem (enlarged hypocotyl), which lends it an unusual appearance for a Brassicaceae plant. Round and convex, kohlrabi has white flesh and green or purple skin, with long stalks of the same color. The plant has broad bluish-green leaves that sometimes seem disproportionately large compared to the size of the bulb.

This delicate, sweet, and fragrant vegetable is a particularly rich source of vitamin C and minerals, especially when eaten raw, either grated or thinly sliced.

Kohlrabi can be diced and baked, deep-fried, or pan-fried. The leaves are best enjoyed when still young and tender and cooked like spinach.

Kohlrabi

"While kohlrabi is better known in Europe as a spring vegetable, I also grow it as a winter crop. It's a big hit when I bring it to market!"

Jean-Martin Fortier

Planting

Seeding in the Spring

Start spring crops in a greenhouse maintained at a constant temperature, between 68 °F and 72 °F (20–22 °C). Use plug flats, sowing 1 seed per cell, then keep the soil moist at all times. Transplant the seedlings once they have developed a strong root ball that fills the cell and holds together well. With this approach, you can plant kohlrabi seedlings over several weeks in order to stagger future harvests as spring kohlrabi grows quickly and is ready for harvest 6 weeks after planting.

Preparing the Soil

Use a broadfork to loosen the bed, then spread a mixture of alfalfa flour (2-0-2) and pelleted chicken manure (5-3-2). Rake the surface, then break up the soil with a power harrow set to a depth of 2 inches (5 cm). There's no need to loosen the soil deeply as kohlrabi has a fairly shallow root system.

Tip from Jean-Martin Fortier

It is very important to water regularly and evenly to avoid cracking and to produce a sweet taste. Plants exposed to excessive heat or dry spells may develop a bitter flavor.

Planting

1 Harden off plants: leave them outdoors, under a row cover, and reduce watering slightly for 7 to 10 days before planting. Mark the rows with a marker roller and drop a seedling at each intersection of the lines drawn on the bed (see p. 42 for spacing).

2 Using a dibber, make a hole and bury the root ball, then pack down the soil with your hands to ensure adequate contact with the roots.

3 Water thoroughly, either by hand, with a gentle spray, or with a drip irrigation or sprinkler system.

Care After Planting

Use floating row covers under tunnels until the end of March. Outdoors, from April onwards, continue using a row cover, but pull it over metal hoops laid out like a low tunnel. Keep the row cover on until outdoor temperatures reach 72°F (23°C), then use the metal hoops to support insect netting.

Summer Seedings

Crops seeded in the summer, for a fall or winter harvest, should be started in July in a tunnel or unheated greenhouse. Use plug flats, keeping the soil moist until germination. As soon as the plants are established, monitor watering and make sure temperatures do not get too high by ventilating the tunnel or greenhouse.

When the root balls can hold together well, after about 4 to 5 weeks, the seedlings are ready to be planted in the field.

Planting

Loosen and amend the beds as you would for spring kohlrabi, then plant the seedlings the same way. The only difference is that fall and storage varieties (Kossak and Superschmelz) produce larger bulbs and require an in-row spacing of 10 to 12 inches (25–30 cm).

Tip from Jean-Martin Fortier

In hot dry summers, regularly water kohlrabi to keep the soil moist at all times. Having a sprinkler or drip irrigation system set up is essential to ensure consistent plant growth and good root development.

Maintenance

Fertilization

Spring kohlrabi is fast growing and shallow rooted. The fertilizer added during transplanting will be enough to meet its needs. However, fall and storage varieties produce larger bulbs that are heavier feeders and require additional fertilizer while growing. Spread chicken manure between rows, after weeding and before watering, which helps dissolved nutrients permeate the soil.

Weeding

The foliage of mature kohlrabi shades the soil, naturally limiting weed growth. It's especially important to watch for weeds in the early stages of the crop, when they are in direct competition. In the first few weeks, cultivate the bed with a collinear hoe or stirrup hoe. Then, when the bulbs are bigger and the blade of the hoe might damage them, switch to using a wire hoe once every 10 to 15 days.

Plant Protection

Like all Brassicaceae, kohlrabi is susceptible to damage by cabbage moths and flea beetles, which make holes in the leaves; cabbage flies, whose larvae damage the bulbs; and gall midges (cecidomyia), which cause plant deformation. Crop rotation can serve as a preventive form of pest control. But, above all, the use of insect netting is highly effective. After transplanting seedlings in the field, set up metal hoops and row covers, which will protect the crops from the cold until the end of April. Then, starting in May, these hoops can support insect netting, which must be kept on until harvest!

Harvest

Lifting the Crop

Kohlrabi is shallow rooted and can be harvested without using a tool to loosen the ground. However, this should be done when the soil is moist and loose, after watering or rain.

Harvest spring varieties when the bulb is fairly small, about 2 to 3 inches (5–7.5 cm). If they grow any bigger, the flesh may be fibrous and less sweet, even bitter. Remove yellow leaves with a knife or pruning shears, keeping healthy, younger green ones to be eaten as cooked greens. With these left on the plant, you can make bunches of 3 to 5 bulbs.

Fall and storage varieties are harvested shortly after the first frosts. Depending on the region, they can remain in the field under a layer of straw, allowing growers to harvest them on demand. Typically, kohlrabi is harvested when it is 6 to 8 inches (15–20 cm) wide and before they begin to elongate. Cut stalks and foliage flush with the bulb, then transfer them into crates to be kept in a storage room.

Tip from Jean-Martin Fortier

Storage kohlrabi can be harvested over the course of 2 weeks, starting with the largest ones. The rest can be harvested later, sometimes after frosts that intensify their flavor.

Storage—Cold Room

When fall and winter kohlrabi come in from the field, dunk them in cold water and shake off any remaining soil. Let the bulbs air-dry, then store in crates kept in a cold room at 36°F (2°C). Under these conditions, they will keep for up to 4 months.

Storage—Root Clamp

Kohlrabi can keep very well in a root clamp. This requires digging a hole near a north-facing wall, in well-drained light soil. Line the pit with fine mesh to prevent rodent damage and add a layer of straw. Next, store roots in plastic crates placed in the pit. To finish, cover the crates with straw and dead leaves. Kohlrabi can be collected on demand. Under these conditions, they should keep until February or March of the following year.

Storage—Cellar

Home gardeners who do not have a cold room can store kohlrabi in a cellar. Strip all foliage from the bulbs before storing them in crates. Maintain temperatures above freezing and ventilate the room, while monitoring the bulbs to remove any that begin to spoil.

Data Sheet

COMMON NAMES: Helianthus, paleleaf woodland sunflower.

SCIENTIFIC NAME: *Helianthus strumosus.*

FAMILY: Asteraceae.

REQUIREMENTS: *Helianthus strumosus* is undemanding and grows in good garden soil that is loose, rich, and well drained.

It thrives in the sun and can tolerate slight shade and periods of drought, which may cause leaves to wilt, but will not impact rhizomes.

SPACING: 1 to 2 rows per bed, with 14 to 16 inches (35–40 cm) between rows, with rhizomes spaced 20 to 24 inches (50–60 cm) apart.

PLANTING: March to April.

DAYS TO MATURITY: 180 to 200 days (nearly 6 months).

ENEMIES: Powdery mildew on foliage. Rodents, which are fond of rhizomes left in the field over winter.

VARIETIES: Only the type species is cultivated. There are no varieties.

NOTE FROM JEAN-MARTIN FORTIER

Cultivate *Helianthus strumosus* apart from your other vegetables as its strong and vigorous growth makes it invasive.

If this crop is not grown in a sheltered area, it requires staking because its long stems bend in the wind.

Helianthus strumosus
(Paleleaf Woodland Sunflower)

A close cousin of Jerusalem artichokes, *Helianthus strumosus* is a widely forgotten perennial crop. It is native to North America and was brought to Europe at the start of the 20th century, well after Jerusalem artichokes, which had crossed the Atlantic in the mid-17th century. This sunflower, in the Asteraceae family, is a decorative plant with stems up to 6 to 10 feet (2–3 m) tall. *Helianthus strumosus* leaves are rough to the touch—differentiating it from Jerusalem artichokes, which have smooth leaves—and are lanceolate and green. Yellow flowers grow in terminal clusters that bloom in late summer and attract bees and butterflies. This plant can become quite invasive over the years, producing rhizomes connected to its collar through a dense network of roots. Unlike Jerusalem artichokes, which are more angular with a pinkish-beige color, *Helianthus strumosus* rhizomes are elongated, with a smooth, light brown skin. Their flesh has an artichoke flavor that is much tastier, subtler, and more distinct. However, the rhizomes of both plants contain inulin, which causes intestinal gas! Young leaves can be cooked, like spinach, while the rhizomes can be baked and eaten in gratins, purées, soups, and even salads.

Helianthus strumosus

> "*Helianthus strumosus* is a culinary curiosity beloved by food lovers and chefs looking for novelty!"
>
> Jean-Martin Fortier

Planting

Dividing Rhizomes

There's no need to buy rhizomes every year as the fragments left in the ground after harvesting easily grow back. This is why we recommend always growing *Helianthus strumosus* in the same spot as the crop naturally regenerates. If you move your plants, make sure no rhizomes remain after harvesting by laying a silage tarp (occultation tarp) over the area for several weeks before starting another crop.

Preparing the Soil

Loosen the bed with a broadfork, then add compost and pelleted chicken manure (5-3-2). Use a rake to remove coarse aggregates from the soil surface and to level the ground.

Tip from Jean-Martin Fortier

Because of its rapid growth, size, and large volume of leaves and stems, *Helianthus strumosus* will quickly deplete the soil, which must be amended before planting then and every subsequent year, if the crop remains in the same spot.

Planting

Mark out the row with a string pulled taut, then use a dibber to make holes every 20 inches (50 cm) for only 1 row per bed, or every 28 inches (70 cm) and staggered for 2 rows. Bury rhizomes under 4 to 6 inches (10–15 cm) of soil. Thoroughly water each plant with a low-flow water supply rather than a spray head, to compact the soil properly around the rhizome. Rooting happens quickly, and young leaves soon sprout without any need for a row cover on the bed.

Care After Planting

Hill plants when the shoots are 8 inches (20 cm) tall to ensure that rhizomes become well anchored. This helps stems to stay upright in windy conditions but does not eliminate the need for stakes that stop them falling into the aisles. Hoe the beds for the first few weeks, then mulch the soil in the summer to keep it cool and avoid watering.

Harvest

When the leaves wilt and turn black after the first frosts, cut the stems down to 16 inches (40 cm). The rhizomes are now ready to be eaten. *Helianthus* is harvested on demand because it is better to leave the rhizomes in the ground to keep them from drying out. The plants can withstand temperatures as low as 5°F to -4°F (-15°C to -20°C). When harvested fresh, this crop is also easier to digest.

Tip from Jean-Martin Fortier

If you need to free up the bed, harvest the crop all at once. Store the rhizomes in crates filled with sand that is kept moist and leave them in a dark, frost-protected cellar. However, this approach will lead to a loss of flavor.

Data Sheet

COMMON NAMES: Turnip, white turnip, salad turnip.

SCIENTIFIC NAME: *Brassica rapa* L. var. *rapifera* (spring turnip) and *Brassica rapa* L. var. *rapa* (fall turnip).

FAMILY: Brassicaceae.

REQUIREMENTS: Turnips are undemanding and grow in all kinds of good garden soil, even with high clay content. Plants can tolerate soils with low nitrogen and organic fertilizer content.

SPACING: 4 rows per bed set 6 to 8 inches (15–20 cm) apart, with a 1.25 to 2 inch (3–5 cm) in-row spacing for spring turnips and 4 to 6 inches (10–15 cm) for fall turnips. With intensive spring crops, you can grow up to 5 rows per bed.

SEEDING: Staggered from March to the end of May (spring harvest) then from late July to late August (fall harvest).

DAYS TO MATURITY: from 35 to 50 days (spring turnips) to 80–90 days (fall turnips).

ENEMIES: Flea beetle, cabbage maggot.

VARIETIES: Hakurei and Tokyo (round turnips), Demi-long de Croissy and Marteau (long turnips), De Milan, Scarlet Queen, and Jaune Boule d'Or (our favorites, for their color).

NOTE FROM JEAN-MARTIN FORTIER

Asian varieties, harvested and eaten like radishes, have undeniable market appeal, for both their appearance and sweet flesh.

Turnips

The term "turnip" refers to spring turnips (*Brassica rapa* L. var. *rapifera*) and fall turnips (*B. rapa* L. var. *rapa*), biennial plants that were known to Greeks and Romans. They grow a storage organ (enlarged hypocotyl) that can be big, in varieties intended for livestock fodder, or small, in varieties widely grown for household consumption until the end of the 18th century. In Europe, turnips were then dethroned by the cultivation and subsequent success of potatoes, brought from South America.

While spring turnips produce a small, round, sometimes elongated root, with white flesh and a white and purple exterior, fall turnips tend to be bigger, with purple skin.

Turnips are rich in minerals and vitamins, especially when eaten raw, and taste similar to cabbage, due to the presence of thioglucosides and oxalic acid. Young leaves can be eaten cooked, like spinach. Low in calories, turnips can have a bitter flavor and suffer from a reputation as a poor man's vegetable, which diminishes their appeal with consumers. Despite this, salad turnips are experiencing a surge in popularity. They can be eaten raw or cooked into a compote, purée, and soup or diced and pan-fried.

Spring Turnips

"Here in Quebec, spring turnips are known as rabioles, a lovely term that is also used in some French regions, such as Limousin!"
Jean-Martin Fortier

Planting

Preparing the Soil

Use a broadfork to loosen the soil, then spread a 1.5-inch (3–4 cm) layer of compost over the surface with a bed preparation rake. If the soil is too lumpy, break it up with a rotary harrow set to a depth of 2 inches (5 cm). There's no need to reach a significant depth as turnips have fairly shallow roots.

Tip from Jean-Martin Fortier

Turnips are a cool-climate crop that does not tolerate hot summer weather. For this reason, seed successions throughout the spring, until the end of May, because crops seeded later will be less tender and more pungent.

Seeding

1 Fill and calibrate your Jang seeder, then walk it down the bed, making sure the seeds are lightly buried (roughly 0.5 inch or 1 cm deep). Drop around 10 seeds per foot (30 seeds per meter).

2 In the event of a late frost, use a row cover to protect the bed. Lay it over metal hoops, which can later be used to support insect netting.

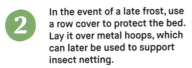

3 About 3 weeks after sowing, when the seedlings are 2 to 2.5 inches (5–6 cm) tall, check the in-row spacing to make sure that they are 1 to 1.5 inches (3–4 cm) apart. Thin as needed.

4 Set up a mini-sprinkler irrigation system or water with a gentle spray. Make sure to keep the soil moist for 10 days after seeding, to promote even germination. Water less frequently afterwards, but the soil should always be kept cool.

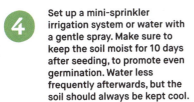

Maintenance

Weeding

Keep the crop weed-free in the weeks after sowing so that weeds will not compete with emerging seedlings. Once the plants have a well-developed true leaf, about 10 to 15 days after germination, cultivate the bed with a flex tine weeder. This also loosens the soil. Subsequent weeding should be done every 12 to 15 days using a stirrup hoe in between the rows.

Plant Protection

Cabbage maggots and flea beetles are fond of turnip leaves, so insect netting is essential, practically from day one. It's therefore a good idea to set up metal hoops right after seeding the crop. They can initially support a row cover, followed by insect netting, which should be fine meshed (0.0138 inch, 0.35 mm), with the edges held firmly in place by bags of sand or soil.

Harvest

Lifting the Crop

Turnips are harvested young, simply pulled out of the ground by hand as they grow right at the surface. This is especially easy after watering or rainfall that loosens the ground. Start by pulling the largest turnips (1.5 inches, 4 cm diameter), leaving room to grow for the smaller ones, which you can harvest the following week.

Storage

Turnips with the tops left on are gathered into bunches of 5 to 7 and will not keep for more than 4 to 5 days in a cold room. They must be sold immediately. Home gardeners can pull turnips as needed, so it's best to sow a few at a time, staggering seeding dates so that you can always harvest small fresh roots.

Fall Turnips

"Fall turnips are quite cold hardy. I sow them late and grow them in tunnels, so they can be harvested even in the middle of winter."

Jean-Martin Fortier

Planting

Preparing the Soil

Fall turnips are often grown after summer crops are finished. This requires you to lay an occultation tarp over the bed to get rid of weeds and plant residue. In 1 to 2 weeks, the tarp will loosen and clean up the soil. Then just finish the surface with a bed preparation rake. If necessary, you can further break up the soil with a power harrow set to a depth of 2 inches (5 cm).

Seeding

Fall turnips are bigger than spring turnips so space rows about 8 inches (20 cm) apart. Use a Jang seeder and make sure seeds are 4 inches (10 cm) apart to allow the turnips to grow larger. When sowing crops towards late summer and early fall, cover beds with a row cover or low tunnel to shelter the seedlings from temperature drops.

Thinning

Once seedlings have a few well-developed leaves, thin by removing those that are growing too close together, leaving them 4 to 5 inches (10–12 cm) apart. Keep the plants that are best-looking and well located within the row.

Maintenance

Weeding

You need to keep this crop clean so that weeds do not compete with young shoots. Use a flex tine weeder until seedlings have 3 or 4 leaves, then use a wire hoe between the rows. This will remove early-stage weeds and loosen the soil surface, allowing irrigation and rainwater to better permeate the bed.

Plant Protection

Although late-summer and fall generations of cabbage flies and flea beetles are less active and less frequent, the risk of infestation still exists. For this reason, we recommend setting up insect netting on metal hoops.

This will also cast a light shade that helps turnips to grow if the end of summer is quite sunny and warm. These hoops will later support a row cover, or even a plastic film, protecting the crop from temperature drops and the first frosts.

Tip from Jean-Martin Fortier

Like spring turnips, fall turnips need moist soil for even germination. This applies to the first seedlings at the end of summer. Later, as soon as the seedlings are well established, switch to watering sparingly, which may not necessarily require setting up mini-sprinklers. Depending on the weather, watering by hand might be enough, but be careful to avoid dry spells that cause water stress and make the harvest more fibrous and tough.

Harvest

When growing fall turnips in the field, harvest them in late October, just after the first frosts, as they cannot remain outdoors unless grown in a tunnel. In loose soil, lift the crop by hand, grasping the foliage to pull out the roots. However, to avoid damaging the turnips, it is best to first loosen the soil with a broadfork or spading fork. Next, cut the tops down to a ½ inch (1 cm) before storing.

Storage

When fall turnips come in from the field, store them in crates or bins in a cold room maintained at 36°F (2°C), where they will keep for 4 to 6 months. Home gardeners can store turnips in cellars, kept in crates filled with sand to stop the roots from drying out. When stored in a dark, cool, and frost-free space, they can keep all winter.

Tip from Jean-Martin Fortier

If you are careful to properly store fall turnips, upright, in sand, and in the dark, you may notice shoots emerging in the winter. They are white and tender, with a sweet delicate taste, and you can eat them in a salad, just like an endive.

Focus on Rutabagas (Swedes)

Belonging to the Brassicaceae family and closely related to rapeseed, this root vegetable has a poor reputation as it was a staple food in World War II, due to its easy cultivation, rapid growth, and high yields. It is rich in sugar, mineral, and vitamin content.

Rutabagas (*Brassica napus* subsp.*napobrassica*), also called swedes, Swedish turnips, neeps, and sometimes simply turnips, are a biennial species derived from a cross between cabbage and turnip, dating back to the Middle Ages. The part we eat is an enlarged hypocotyl, a swelling between the stem and the root system. Yellow-fleshed varieties are grown for household consumption, while white-fleshed varieties are used as livestock fodder.

This vegetable has a strong flavor, so it needs to be cooked to obtain a milder taste. It can be used in a stew, purée, or soup; can be stuffed; and can also be eaten raw, grated or in a carpaccio.

Rutabagas are undemanding and grow well in all types of soil, especially cool soils. Sow into trays in March, then plant seedlings in the field in April, with a 14-inch (35 cm) spacing in all directions. Begin harvesting in mid-June. Direct seed rutabagas from May to June, with the same spacings, for demand harvesting from September to November.

Data Sheet

COMMON NAMES: Oca, Peruvian oca, also called yam in New Zealand.
SCIENTIFIC NAME: *Oxalis tuberosa.*
FAMILY: Oxalidaceae.
REQUIREMENTS: Oca grows well in rich, loose, light, well-drained good garden soil, in sunny and warm locations.
SPACING: 1 row down the middle of the bed, with tubers set 14 inches (35 cm) apart.

PLANTING: April to May.
DAYS TO MATURITY: More than 200 days (6 to 7 months).
ENEMIES: Like its cousin, clover, oca is a naturally dominant plant. It has no specific enemies aside from rodents.
VARIETIES: In the many varieties, the most significant difference is the skin color.

NOTE FROM JEAN-MARTIN FORTIER
Varieties with dark red or bluish-purple, almost black, skin have the least acidic flesh. A lengthy cooking time reduces the acidity to reveal sweet and subtle flavors.

Oca (Oca Tubers)

This tuber is native to South America and was brought to Europe in the 19th century, where it was so popular and cultivated to such an extent that it was expected to rival potatoes. In the Andes, it was eaten as much, if not more than, potatoes. However, oca was not as successful as anticipated and fell into obscurity a few decades later at the start of the 20th century. Today, its cultivation and consumption is fairly limited; oca can be found in original, gourmet meals prepared by some chefs and in the kitchens of a few home gardeners who love unusual heirloom vegetables.

Grown as an annual in market gardens and home vegetable plots, oca is a perennial belonging to the clover family. It forms a 6- to 15-inch (15–40 cm) bush and produces seedless yellow flowers. The leaves look identical to clover and grow from short, cylindrical, ribbed tubers, with light pink, yellow, red, and purple skin. Oca flesh, quite similar to that of potatoes, has a floury texture, with a slightly sweet taste and hints of sorrel. It is rich in oxalic acid, which lends it some acidity, but that disappears within a few days of exposure to light after harvest.

Oca can be steamed, sautéed, or simmered in a stew.

Oca

—————●—————

"You might worry that oca could take over your garden like its clover cousin, but the tuber is destroyed by frost and won't propagate from one year to the next!"
Jean-Martin Fortier

Planting

Purchasing Tubers

Since oca plants do not produce seeds, tubers are their only means of propagation. Tubers can be purchased at the end of the winter, but you can also save some in the fall, after the harvest is over, and plant them the following spring. As you would with potatoes, allow oca tubers to sprout by putting them in crates left in a warm and bright space, either in a greenhouse or on a veranda. Plant pre-sprouted tubers in open ground as soon as the risk of frost has passed.

Preparing the Soil

Use a broadfork to loosen the bed, then spread compost and pelleted chicken manure (5-3-2).

Next, use a bed preparation rake to incorporate the amendments into the soil surface. This will also break down coarser aggregates.

Tip from Jean-Martin Fortier

It is easy to prepare oca tuber cuttings. Using a sharp knife, simply cut a tuber into several slices, keeping 1 or 2 buds on each one. Then, transplant them into a flat filled with a mix of sand and potting soil. Tubers kept under a plastic or glass cover and exposed to artificial light or indirect sunlight will take root quickly.

Planting

In cold climates, plant pre-sprouted tubers in open ground around mid-May. Place each one in a hole made with a dibber, then compact the soil around the tuber and water thoroughly with a gentle spray. In mild climates, plant tubers at the end of April, covered with roughly 3 inches (7–8 cm) of soil, and then water.

Care After Planting

Use floating row covers to mitigate temperature variations between day and night until the end of May. Water regularly after leaves appear and continue throughout the season but avoid overwatering. When the stems are 6 to 8 inches (15–20 cm) tall, use a hoe to hill the row and promote tuberization along the stems. Cultivate along the row and in-row until the oca crop covers the surface and mulch the bed to retain moisture and improve tuberization.

Tip from Jean-Martin Fortier

In summer and in full daylight, oca leaves droop once it gets hot then return to a normal position at the end of the day and overnight. This allows the plant to protect its foliage from the scorching sun and does not necessarily indicate a lack of water.

Harvest

Tubers form in late summer, towards the end of September, as the days get shorter. At this stage, the crop must be hilled and covered with a black plastic film, supported by metal hoops or laid over beds. This will amplify the tuberization process and protect the crop from frosts. Harvest oca gradually until the end of October for cold climates and until the end of November for mild climates.

Lifting the Crop

When cold weather and the first frosts turn the foliage brown, use a broadfork or spading fork to lift the crop so you can dig up the tubers. Brush off soil residue and remove root and stem fragments. If you leave oca tubers on the ground and exposed to the sun for about 10 days, while ensuring that they do not freeze, they will lose some of their acidity.

Storage

Store clean and dry tubers in a dark and frost-free space like a cold room or a cellar, as with potatoes. You can mix them with moist sand to slow the dehydration process, allowing them to keep for 3 to 5 months.

Focus on Mashua

This climbing plant belonging to the Tropaeolaceae family is native to South America. In the 19th century, it was brought to Europe to replace potatoes that at the time had been decimated by late blight. Mashua is closely related to garden nasturtiums, a plant with spicy edible flowers, but differs by growing plump, bumpy, tapered tubers underground. These are quite starchy, but steaming reduces this and reveals subtle flavors of chestnut and artichoke mixed with fennel and almond.

You can buy tubers or save some of the harvest from one year to the next. Plant them in April and May—after the risk of frost has passed—into rich, loose, deep soil sheltered from wind and intense heat. They grow particularly well in light shade. Bury tubers 5 inches (12 cm) deep with a 32-inch (80 cm) in-row spacing. When shoots appear, hill the plants and set up stakes to train their twining stems.

As with oca, tubers form late, around the end of September, which means that a row cover should be set up to protect the plants from early frosts and keep them in the field as long as possible.

Harvest tubers once the foliage is entirely brown and frost damaged. When mixed with sand and stored in crates, they will keep for several months in a dark and frost-free cellar.

Data Sheet

COMMON NAME: Parsnip.
SCIENTIFIC NAME: *Pastinaca sativa*.
FAMILY: Apiaceae.
REQUIREMENTS: Parsnips are easy to grow and very cold hardy. When grown in fertile, loose, deep soil, they can develop consistent, uniform large roots.
SPACING: 3 rows per bed set 8 to 10 inches (20–25 cm) apart, with a row spacing of about 4 inches (10 cm).

SEEDING: From mid-March to late May.
DAYS TO MATURITY: Approximately 150 days (5 months) after seeding.
ENEMIES: Carrot flies (*Psila rosae*), nematodes, and wireworms, powdery mildew, and late blight affect the foliage, and, of course, rodents are a threat to all root vegetables!
VARIETIES: Parsnips come in all shapes and sizes, from short and broad to long and tapered. Guernsey Half Long (a classic, the best and most suitable for cooking!), Hablange Weisse (white, tapered, cold-hardy), Turga (sweet taste), and Dagger.

NOTE FROM JEAN-MARTIN FORTIER
Although parsnips are slow-growing and occupy a bed for several months, they have an excellent shelf life, which makes them a must-have root vegetable that will complement other winter crops.

Parsnips

Parsnips—often mistaken for white carrots, and at times for root parsley and turnip-rooted chervil, which belong to the same family—are a popular root vegetable with a long history. They were initially cultivated as an easy-to-grow, nourishing food crop until dethroned by potatoes and carrots in the 19th century.

Native to Central and Southern Europe, they grow wild everywhere, particularly along embankments, in open fields, and in ditches with clay soil. In the first year, this biennial species produces a root and leaves similar to flat parsley, then it flowers the following year. The tapered roots have thick, tough, yellowish-beige skin that protects the fibrous white flesh. Due to a high volatile oil content, they have a pungent odor that is reminiscent of parsley and carrots. When harvested after the first frost, this scent is intensified, at which point it is best to cook them, preferably combined with other vegetables to temper their flavor. Parsnips are rich in fiber, protein, and a much higher vitamin C content than carrots.

In the United Kingdom, Canada, and the United States, parsnips have continued to be a popular crop, where it is featured raw—in salads, like celery—or cooked in stews, purées, soups, or mixed vegetable dishes.

Parsnips

"A favorite in Europe and North America for for centuries,
parsnips rightly deserve a place on our plates!"
Jean-Martin Fortier

Planting

Because the germination process can be
delicate and difficult, you should only use
fresh seeds from the previous year. These
can be purchased or harvested from a few
plants that were allowed to go to seed.
Before sowing, careful soil preparation is
paramount. It's best to seed rather densely,
to ensure even germination, and thin later.

Preparing the Soil

Use a broadfork to loosen your beds, then
apply compost. Next, run a power harrow
down the bed at a depth of 2 inches (5 cm)
and rake the surface to remove any clods
and coarse aggregates. The seedbed must be
prepared carefully to a depth of roughly
1 inch (2–3 cm).

Tip from Jean-Martin Fortier

Parsnips can be quite slow to germinate and may not emerge for 15
or 20 days! You must therefore be patient and, above all, pay close
attention to growing conditions in the days following the seeding.

Seeding in the Field

1 Fill and calibrate your Jang seeder and start seeding 3 rows, dropping about 10 seeds per linear foot (30 seeds/meter) into each furrow.

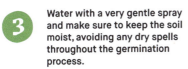

2 After seeding, cover the bed with a mixture of sand and fine soil and then tamp it again with a seedbed roller. Perfect contact between the seeds and the soil will improve germination.

3 Water with a very gentle spray and make sure to keep the soil moist, avoiding any dry spells throughout the germination process.

4 Cover the bed with a floating row cover to prevent the soil from drying out and maintain a constant temperature, which is an essential condition for germination.

Thinning

When parsnip shoots are about 4 inches (10 cm) tall, the row must be thinned out, spacing them 4 to 6 inches (10–15 cm) apart. Select the best-developed and most vigorous seedlings and hand pull the weakest ones. After thinning, run a flex tine weeder down the bed to loosen and level the soil surface, then replace the row cover.

Maintenance

Weeding

Because parsnips are so similar to carrots, their maintenance practices are alike, starting with weeding. Run a flex tine weeder down the bed once seedlings have 2 or 3 leaves, then repeat every 12 to 15 days for the first 2 months. This is recommended to uproot weeds and loosen the soil.

When the parsnip foliage is well developed and spilling out beyond the row, run a hoe with bio-discs down the bed a few times to eliminate weeds and to gently hill the tops of the roots, which will reduce carrot fly damage.

Any remaining weeds have to be removed by hand.

Irrigation

Regular watering is required at the beginning, after which parsnips adapt to summer rainfall. Mulching the bed with bark, leaves, straw, or wood chips helps keep the soil cool and reduces the need to irrigate.

Plant Protection

To avoid carrot fly damage, we recommend setting up insect netting supported by metal hoops in the late summer when this parasite's final nuptial flights occur. Note that the carrot fly will only be present if a few carrot beds are near the parsnips!

Harvest

Parsnips are root vegetables that growers harvest from the end of October. Since they are cold hardy, you need not lift them all at once as they can overwinter in the ground. With this approach, the bed will remain occupied by parsnips, so it may be wise to evaluate the cost of monopolizing this space over long winter months compared to planting a winter crop.

Lifting the Crop

To make root extraction easier in the dead of winter, especially when the ground is frozen, it's a good idea to cover the soil with dead leaves, straw, or wood chips. This mulch will insulate the soil and make it easier to use a broadfork or spading fork when lifting the roots.

Storage

If you are harvesting the entire crop, just after the first frosts, pull the parsnips and wash off any remaining soil, then sort them so that you store only roots that are in good condition and showing no signs of damage. Cut the foliage down to roughly ½ inch (1 cm) and put the roots into harvest crates. Market gardeners should store parsnips in a cold room kept at 36°F (2°C), while home gardeners can use the same crates, but filled with wet sand and then stored in a dark and frost-protected cellar. Parsnips can keep this way for 2 to 3 months.

Data Sheet

COMMON NAMES: Sweet potato. In North America, this plant is also sometimes referred to as a yam, though it is not a true yam.

SCIENTIFIC NAME: *Ipomoea batatas.*

FAMILY: Convolvulaceae.

REQUIREMENTS: Sweet potatoes need a sunny, warm climate (72–75°F, 22–25°C, on average). They grow well in sandy, loose, rich, deep soil that warms up quickly in the spring, which promotes the development of large tubers.

SPACING: A single row down the center of the bed, with a 12-inch (30 cm) in-row spacing.

PLANTING: Late April to early May.

DAYS TO MATURITY: 180 days (6 to 7 months), including a harvest period spread over 1 month.

ENEMIES: Fusarium wilt causes black spots on tubers during their growth and in storage. Late blight and powdery mildew damage foliage in the summer, as do mites, aphids, and whiteflies when the crop is grown indoors.

VARIETIES: Beauregard (type species), Bonita (early, high-yielding), Murasaki (less sweet, with a hazelnut smell), Evangeline (distinguished by orange flesh and sweet taste).

NOTE FROM JEAN-MARTIN FORTIER

Sweet potatoes take up a lot of space due to their expansive foliage. For market gardeners, this root vegetable is an alternative to potatoes, but it requires space and time and may not necessarily be a priority.

Sweet Potatoes

Although the sweet potato is a perennial species, market gardeners grow it as an annual crop. It belongs to the Convolvulaceae family and is related to bindweed, a plant feared by gardeners. Native to South America, it is widely consumed in China, Papua New Guinea, and East Africa. In Europe, ornamental varieties are most common, referred to as morning glory, and grown for their colorful foliage. In recent years, however, this elongated tuber—ranging in color from yellowish-brown to pink or purplish-red—has increasingly been grown as a food crop.

Sweet potato tubers have a white, yellow, orange, or purple flesh that is rich in starch, carotene, and vitamin B6, as well as minerals and sugars. They produce climbing vines that creep along the ground. The foliage is lush, deciduous, and green in the edible varieties, with large trumpet-shaped flowers that are identical to bindweed blooms. New seedlings can be grown from tuber cuttings, which easily sprout roots and shoots, or by saving tubers from one year to the next, much like seed potatoes.

The edible leaves can be cooked like spinach, but this crop is mainly grown for its tubers, which are prepared like potatoes, in purées, gratins, soups, and even cakes and pies because of their mild, sweet flavor.

Sweet Potatoes

———————•———————

"Despite its humble origins, the sweet potato remains a cherished crop, beloved for its rich flavor and nourishing qualities by food lovers everywhere."
Jean-Martin Fortier

Planting

Growing

For market gardeners and home gardeners, purchasing tubers or seedlings grown from cuttings—some cultivars are exclusively sold by the crop breeders—is the most practical way to start a crop. Type species tubers can also be saved from one year to the next, provided they are harvested in the fall from healthy virus-free plants and have not suffered any damage during overwintering.

Germination

Sweet potatoes require heat and plenty of light to trigger sprouting and growth. We recommend preparing slips, which are rooted sprouts produced by the tubers. Place tubers side by side, on a layer of potting mix, in a box left in a bright warm space. Use a heating mat to provide the temperatures (72–77 °F, 22–25 °C) needed for germination. In this emergence phase, spray the tubers with warm water to keep the substrate moist and to stop them from drying out.

Preparing the Soil

Loosen the bed with a broadfork, then spread and incorporate a thick layer of compost. Tubers require well-drained soil, so it's best to create a raised mound that is roughly 8 inches (20 cm) high, to add to the depth already provided by the raised bed.

Planting

Plant slips (see illustration below) or tubers after making holes with a hand trowel every 12 to 16 inches (30–40 cm). Place the sprouted tuber or seedling in the hole, then carefully fill it in. Water around each one to help compact the soil.

Care After Planting

Install a row cover or plastic film over metal hoops to avoid damaging young shoots. By creating a closed and protected environment that is sheltered from the wind, you can maintain a constant temperature, which is conducive to new stem production.

Tip from Jean-Martin Fortier

Since the tubers are quite bulky, keep only the smallest ones for next year's spring planting. If you don't have enough tubers or need more plants, cut a few lengthwise and place them on a layer of potting mix. After the dormant buds produce shoots, simply cut the tuber slice crosswise to get multiple pieces, and thus several plants.

Maintenance

Weeding

In the weeks following planting, hoe the bed to control weeds while also loosening the soil. Soon the crop's twining stems will grow to about 15 feet (4 or 5 m) long, and the foliage will cover the bed surface, preventing weed growth and eliminating the need to hoe. At this point, you can remove the row cover or plastic film and the metal hoops.

Irrigation

Although sweet potatoes grow in warm and even arid climates and can tolerate poor soils, regular watering will promote tuber growth for improved yields. This is why drip irrigation should be set up in May, with a line of drip tape running along the base of each plant. Towards the end of summer, limit watering as an excess may cause tubers to burst and rot.

Tip from Jean-Martin Fortier

From mid-July onwards, you can promote tuber growth by withholding water for 2 to 3 weeks, encouraging the root system to grow deeper into the soil as it searches for cooler temperatures and moisture. During this time, tubers will also lengthen and divide further.

Harvest

Tuber growth, called tuberization, speeds up as day length decreases (late August), and optimum size is reached at the end of October. Therefore, you should, if possible, wait until November to begin harvesting. In cold climates, tubers must be gathered before the first hard frost. Yellowing foliage is often a sign that it is time to harvest. Typically, this occurs sometime in October.

Lifting the Crop

Lightly loosen the soil with a broadfork, being careful not to cut the tubers. Harvest by hand, gently pulling on the stems. Place tubers in crates and keep them in a warm sheltered space like a tunnel or dry room for 2 to 4 weeks. This curing step is essential to initiate the healing process, sealing wounds where tubers were attached to roots or stems.

Storage

Unlike other root vegetables, sweet potatoes are stored in crates without being buried in sand. Keep them in a dark ventilated room maintained at higher temperatures than average, 50°F to 57°F (10–14°C). Under these conditions, the tubers will keep for up to 4 or 5 months and require constant monitoring to remove any showing signs of decay.

Data Sheet

COMMON NAMES: Root parsley, parsley root, Hamburg parsley, turnip-rooted parsley.

SCIENTIFIC NAME: *Petroselinum crispum* var. *radicosum.*

FAMILY: Apiaceae.

REQUIREMENTS: Root parsley is easy to grow and very cold hardy.
It prefers cool, loose, deep, and fertile soil that allows the plant to grow long, thick, tapering roots.

SPACING: 3 rows per bed, set 8 to 10 inches (20–25 cm) apart, with an in-row spacing of about 4 inches (15 cm).

SEEDING: From mid-March to late April.

DAYS TO MATURITY: From 150 to more than 270 days if the roots are kept in the ground over winter.

ENEMIES: Carrot fly (*Psila rosae*), spots that appear on roots left in the field,

and—as with all root vegetables—rodents!

VARIETIES: Halblange (early variety, not very sensitive to frost), Fakir (half-long root, smooth), Hamburg.

NOTE FROM JEAN-MARTIN FORTIER Although root parsley is a slow-growing vegetable that occupies the field for many months, it is a storage vegetable that complements other winter crops.

Root Parsley

Wild parsley is found throughout the Mediterranean region where cultivation initially began. It later spread to the rest of Europe and is now grown as an aromatic. The most well-known types of parsley, annual and biennial, are flat parsley and curly parsley that are grown for their edible foliage. One variety is grown for its root, mainly in Central and Northeastern Europe: root parsley.

It produces a very white taproot—which distinguishes it from the closely related parsnip—that is short, thick, fleshy, and sometimes round. The root, which tastes a bit like celeriac and parsnip, is rich in vitamins A, B, and C, and many trace minerals.

Parsley root can be eaten raw and grated; cooked, in mashes, soups, and gratins; roasted in the oven, or added to a mixed vegetable dish, with potatoes and carrots. The flat green foliage is similar to flat-leaf parsley and can be eaten fresh, used as garnish on salads, or added to baked dishes.

Root Parsley

"More common in Europe than in North America, root parsley is often seen as a perennial type of parsley that can be grown for its foliage."
Jean-Martin Fortier

Planting

With root parsley, the germination stage is delicate and difficult but can be improved by soaking seeds overnight in a glass of water and carefully preparing the soil before seeding.

Preparing the Soil

Use a broadfork to loosen the soil, then spread a 2-inch (5 cm) layer of compost over the bed and mix it in with a power harrow. Next, rake the bed to further break up the surface, maintaining a depth of about 1 inch (2–3 cm).

Seeding in the Field

 1 Fill and calibrate your Jang seeder and start seeding 3 rows, at 1- to 2-inch (2.5–5 cm) spacing, or seed 1 row at a time in 3 passes.

2 Cover the bed surface with a mixture of sand and soil then tamp it down with a seedbed roller to ensure there is good contact with the seeds.

3 Water the bed with a gentle spray, then set up a row cover to keep the soil from drying out and maintain a constant temperature, which is an essential condition for germination.

Thinning

Once the seedlings are about 4 inches (10 cm) tall, thin the row to achieve a 6-inch (15 cm) spacing, keeping the strongest and best developed plants.

Maintenance

Weeding

Run a flex tine weeder down the bed as soon as the seedlings have 2 or 3 leaves, then repeat every 12 to 15 days for the first 2 months. When the foliage is big enough and overflowing into the aisles, hoe between the rows to remove weeds and loosen the soil.

Irrigation

Root parsley needs soil that is kept cool, even moist, which requires regular watering, particularly during hot and dry spells. By mulching the soil with bark, leaves, straw, or wood chips, you can save on watering.

Plant Protection

To prevent foliar damage caused by carrot flies—although rarely seen in root parsley—plan to set up insect netting supported by metal hoops when summer begins.

Harvest

Root parsley can be harvested as early as September, 4 to 5 months after seeding, but it is so cold hardy that it can overwinter in the field and be harvested on demand. This means the bed will be occupied for 8 to 9 months. You will need to weigh the value of monopolizing this bed against potential gains from planting other late-winter crops, like early vegetables.

Lifting the Crop

The roots, which are sometimes fragile and brittle, can be pulled after the soil has been loosened with a broadfork or spading fork. If you are harvesting the crop on demand over the winter, mulching the soil with dead leaves will provide frost protection and make uprooting easier.

Storage

When harvesting in September or October, cut the tops down to ½ inch (1 cm), clean the roots to remove soil residue, and sort them. Only store roots that are in good shape, showing no signs of damage or decay, into harvest crates. Market gardeners can keep them in a cold room maintained at 36°F (2°C).

Home gardeners can store the harvest in crates filled with wet sand and kept in a cellar, in a dark and frost-protected space. Under these conditions, root parsley can keep for 3 to 4 months.

Focus on Turnip-Rooted Chervil

This biennial plant, also known as *Chaerophyllum bulbosum*, belongs to the Apiaceae family and is native to Eastern Europe and Siberia. It is closely related to root parsley but can be distinguished by its stubby, conical roots that are sometimes rounded and beige.

These floury roots are fragrant and slightly sweet, with a flavor resembling chestnuts and potatoes. The foliage, unlike all other types of parsley, is toxic and therefore unfit for consumption.

Turnip-rooted chervil is slow growing. Field crops are seeded in the fall (September–October), while crops grown in tunnels are seeded in the winter (February–March). Sow rows 12 inches (30 cm) apart. A subsequent thinning is required, leaving plants 2 to 2.5 inches (5–6 cm) apart. Water sparingly and cultivate frequently, especially in the spring when weeds grow back.

Harvest the roots from mid-June to early fall when the foliage begins to yellow and fully dry out. Leave harvested roots on the beds for a few days before storing them in crates in a cellar.

The roots develop their full flavor only after several months of storage. Root chervil is eaten steamed—although just lightly as it becomes floury when overcooked—a process that brings out delicate fragrances.

Data Sheet

COMMON NAME: Potato. This includes baby or new potatoes, and storage potatoes.

SCIENTIFIC NAME: *Solanum tuberosum.*

FAMILY: Solanaceae.

REQUIREMENTS: Storage potatoes grow well anywhere, while new potatoes are best suited to areas with mild and frost-free springs as tubers freeze at 32 °F (0 °C). New potatoes thrive in loose, deep, fertile, well-drained light soils.

SPACING: 2 rows per bed with an in-row spacing of about 8 inches (20 cm) for early crops. One single row with an 8-inch (20 cm) in-row spacing for storage potatoes.

PLANTING: From February to March for early crops, from late March to late May for storage varieties.

DAYS TO MATURITY: 80 days for new potatoes and 110 to 150+ days for storage potatoes.

ENEMIES: Late blight *(Phytophthora infestans)*, fungi *(Rhizoctonia solani)* in the germination stage, silver scurf *(Helminthosporium solani)* in storage, the Colorado potato beetle, wireworms (click beetle), aphids.

VARIETIES: There are many; your choice depends on the yields you need, how the potatoes will be cooked (fried, mashed, steamed...), or the color of the flesh, as highlighted by Bleu d'Artois. For early harvests, I like Margod, Charlotte, and Colomba for their yellow flesh (new potatoes); and for storage, Marabel.

NOTE FROM JEAN-MARTIN FORTIER

These days, disease resistance is fast evolving and should be one of the most important selection criteria for organic market gardeners. To limit the impact of diseases, we recommend a long-term rotation of 3 to 4 years between crops.

Potatoes

The potato is a perennial species native to Chile and Peru. Although brought to Europe in the 16th century, it was originally relegated to the status of botanical curiosity. It wasn't until the 17th century that the potato graduated from ornamental plant to food plant, initially only in Germany, Austria, Switzerland, and a part of France. In the rest of France and Europe, this crop did not take off until the 18th century, when the famous agronomist Antoine Parmentier promoted and developed potato cultivation as a solution to the famines of that era. Potatoes were primarily introduced to North America by European settlers.

Potato plants form a herbaceous bush, 20 to 30 inches (50–70 cm) tall, with green leaves and purple flower clusters. Underground, the roots produce swellings, tubers that range in size and shape—including round, oval, and tapered—with pale yellow or blue flesh and brown, red, or bluish-purple skin.

In our diets, potatoes are a source of carbohydrates, protein, and vitamins. They should never be eaten raw because potato organs (skin, unripe fruit, leaf, and sprouts) contain solanine, a toxin. They can be cooked and served in many ways—steamed, baked, roasted, fried, boiled, mashed, and in salads.

New Potatoes

———————————

"New potatoes provide lower yields, but they are tastier for the consumer and more profitable for market gardeners."
Jean-Martin Fortier

Planting

Preparing the Soil

New potatoes grow well in loose, deep, and well-drained light soils that warm up quickly because tubers are planted when soil temperatures are around 50°F (10°C). First, open up the bed with a broadfork, then level the surface with a power harrow and bed preparation rake, for a soft, loose seedbed.

Pre-Sprouting (Chitting)

You should ideally buy new, certified, and disease-free seed potatoes. Two weeks before planting, place them in crates in a dark room kept at roughly 50°F (10°C) and 60–70% humidity, to encourage them to produce sprouts (eyes). Once the sprouts are about ¾ inch (2 cm) long, they are ready to be planted.

Planting

1 Use a shovel or garden hoe to dig 2 trenches in the bed about 5 inches (12 cm) deep.

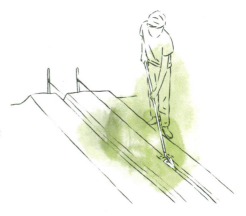

2 Place seed potatoes in the trench, by hand, 8 inches (20 cm) apart. Keep the eyes (sprouts) pointing upwards, taking care not to break them.

3 Use a rake to fill in the trench and cover the seeds with soil, then water thoroughly. For the following week, keep the ground moist to stimulate growth.

Tip from Jean-Martin Fortier

To promote growth and early development, install a low tunnel over the bed or lay a floating row cover over several beds.

Maintenance

Weeding
Remove weeds from the very beginning by running a 5-inch (13.5 cm) stirrup hoe along each side of the row.

Hilling
Once the stems are 4 to 6 inches (10–15 cm) tall, hill the row by mounding soil around the base of the plants, taking care not to bury the leaves.
Repeat this process about 3 to 4 weeks later, when the plants are 12 inches (30 cm) tall.

Plant Protection
Plants are sensitive to the Colorado potato beetle, whose larvae devour their aboveground parts. You can act preventively by introducing host plants (yarrow, nasturtium, and nectar plants) nearby to attract predatory insects (ladybug, ground beetle, green lacewing, etc.) and by installing insect netting or row covers at the start of the season. If pest pressure is strong and you have few plants, pick off larvae and adults by hand. If the crop is larger, a biopesticide application is another option.

Late blight is a highly destructive disease, but you can limit its impacts by purchasing new seeds every year, choosing disease-resistant varieties, and keeping foliage dry through drip irrigation.

Harvest

New potatoes are harvested 7 to 8 weeks after planting, and a few weeks after flowering. This should be done in dry weather, when the soil is moist but not overly wet. Loosen the soil with a broadfork, then pull the stem while shaking it, to extract the tubers. Brush off the dirt and place them in harvest crates.

Storage

Next, soak the tubers in a tub, then hose them down to remove dirt without damaging the skin. New potatoes will keep in a cold room at 40°F (4°C) or in a cellar. However, they should ideally be eaten fresh, in the days just after the harvest, to enjoy their full flavor.

Tip from Jean-Martin Fortier

Hilling potatoes builds a protective layer around the tubers. It shields them from late blight by directing rainwater or irrigation water towards the aisles, preventing spores from reaching the plants.

Storage Potatoes

———————●———————

"With storage potatoes, you are guaranteed to have stock available all winter long. This is an exceptional vegetable!"
Jean-Martin Fortier

Planting

Preparing the Soil

Soil preparation is identical for storage potatoes and new potatoes. Make sure to work meticulously as the crop is in the ground for over 5 months! Because storage potatoes are heavier feeders, you should add mature compost and a mixture of alfalfa meal (2-0-2) or pelleted chicken manure (5-3-2) to the trench before planting.

Planting

Use certified seed potatoes and pre-sprout them 2 weeks before planting. Following the steps for new potatoes, plant storage potatoes in a single row per bed. You may opt to spread pelleted chicken manure down the row before filling the trench. To finish, cover the base of the shoots with soil using a hoe to create a small mound.

Maintenance

Hilling and Irrigation

About 3 weeks after the first hilling, once the plants are 12 inches (30 cm) tall, hill the crop again, weed the beds, and monitor for pests and diseases. Keep a close eye on water inputs as potatoes are prone to water stress.

The plants have an especially high water requirement in the tuber formation (tuberization) stage, which occurs when flower buds appear. Although sprinklers wet the foliage and can promote late blight, they do make it easier to manage watering. However, drip irrigation is preferable as it keeps the foliage dry. This requires setting up the drip tape in the planting stage and burying it so as not to impede the hilling process.

Tip from Jean-Martin Fortier

When storage potatoes reach an adequate size, the plants must be defoliated. This operation involves cutting back the foliage, which triggers a skin thickening process. The crop will be ready for harvest a few weeks later. However, if there is no rush to harvest, you can wait for the foliage to dry naturally in the fall.

Harvest

Storage potatoes are harvested in the fall, 5 to 6 months after planting and 2 to 3 weeks after defoliating the crop. Since the foliage has dried out or been cut back, the absence of stems allows growers to use a broadfork or spading fork to expose the tubers. Use your hands to collect any that may be hidden, then put them in harvest crates. Do not wash them as it would reduce their shelf life. At this stage, remove any damaged, diseased, or green tubers. Lastly, if the soil was wet while harvesting, let the tubers air-dry for a few hours in a ventilated and dark room to avoid greening.

Storage

Keep storage potatoes, unwashed and in crates, in a ventilated and dark room maintained at roughly 50°F (10°C) and 90% humidity.

A cool room is the ideal space to achieve these conditions, which allow tuber wounds to heal. This 1- to 2-week stage is a prerequisite for successful storage. Next, move the crates into a cold room with the same environmental conditions, but with temperatures ranging from 40 °F to 45°F (4–7°C). For home gardeners, a cellar can provide the same conditions.

Tip from Jean-Martin Fortier

Note that potatoes stored below 46°F (8°C) undergo a process that converts starches into sugars, and when temperatures are below 40°F (4°C), this process accelerates, thus limiting shelf life.

Focus on Yacon

This root vegetable is native to Peru and has nothing in common with the potato, except for a similar underground organ and the fact that its French name, poire de terre (meaning "ground pear"), is similar to the French word for potato! At our latitudes, typically above 43°N, yacon is not a hardy plant, though it does have perennial tubers similar to those of dahlias and develops foliage that is considered annual.

Yacon *(Polymnia sonchifolia)* belongs to the Asteraceae family and develops organs that are irregular in shape, with a decent size, dark skin, and a grainy, crunchy flesh—a bit like an almost-ripe pear— that has little flavor. It is eaten raw, cooked, as a side vegetable or in a dessert.

Some tubers can be saved for propagation in the following season. Plant them in May, after the last frosts, in a loose, rich, deep soil. After 1 month, hill the crop to help anchor and stabilize the plants. Mulching these beds is essential to keep the soil cool in the summer and provide frost protection in the fall. This crop has to undergo a tuberization process, which triggers the tuber formation, increasing both their quantity and size.

Lift yacons after the foliage has been killed off by frost. Tubers can be stored unwashed, in a dark, frost-protected cellar for 4 to 5 months.

Data Sheet

COMMON NAMES: Spring radish, also called summer radish or small radish. Winter radish.

SCIENTIFIC NAME: *Raphanus sativus*.

FAMILY: Brassicaceae.

REQUIREMENTS: All radishes grow best in a bright, sunny spot. They need soil that is well drained and loose, especially at depth, as compaction can cause root malformations.

SPACING: Up to 12 rows per bed with an in-row spacing of roughly 1.25 inches (3 cm) for spring radishes, and 3 rows with an 8-inch (20 cm) in-row spacing for winter radishes.

SEEDING: March to September for spring radish, and mid-July to mid-August for winter varieties.

DAYS TO MATURITY: 20 to 30 days for spring radishes and 70 to 90 days for winter radishes.

PESTS: Flea beetle, cabbage maggot, early blight, soil-borne diseases (fusarium wilt, etc.).

VARIETIES: There are so many, but Raxe and Pink Beauty are great little round radishes, and French Breakfast is a nice medium-length, oblong variety. Nero Tondo, Rond Noir, and Noir Long Maraîcher are perfect winter radishes.

NOTE FROM JEAN-MARTIN FORTIER

For professional market gardeners, radishes are an eye-catching and highly profitable crop. They are also quite easy to grow successfully at home, especially when introducing young children to vegetable gardening; they can be harvested soon after seeding, which makes them particularly rewarding.

Radishes

The radish genus includes spring radishes and winter radishes, both biennial species. While spring radishes have been cultivated since the 16th century, winter radishes were widely cultivated in Europe as early as the Middle Ages, and known to Greeks and Egyptians.

Spring radishes are typically 4 to 6 inches (10–15 cm) long, with green leaves growing from a swelling at the base of the stem (hypocotyl). They can be round or oblong, with a red, red and white, or white skin. The flesh—white, crisp, and rich in minerals, sugars, and vitamins—can sometimes be quite pungent. They can be eaten raw, with a pinch of salt, but the root and baby greens can also be cooked.

Winter radishes form a cylindrical taproot that is 6 to 12 inches (15–30 cm) long. The skin is often thick, and can be white, purplish red, or black. It protects the flesh, which tends to be spicy, with a strong flavor reminiscent of mustard due to the presence of a volatile oil (allyl isothiocyanate). Their large green leaves grow in a rosette from which a flowering stem emerges if the root is not harvested. Winter radishes have beneficial digestive properties and provide a nutritional supplement with a high vitamin and mineral content. They are eaten raw, as crudités, or with a pinch of salt.

Spring Radishes

"Radishes are one of the few vegetables I intercrop near
plants that are slower to become established,
like zucchinis, cucumbers, and peas."
Jean-Martin Fortier

Planting

Preparing the Soil
Radishes require a seedbed that is
meticulously prepared at the surface and
well drained at depth as they are sensitive to
standing water after rainfall or irrigating.

For this reason, you can adjust the angle of
the bed surface to make it slope towards the
paths and avoid pooling water.
Loosen the bed with a broadfork and a
power harrow set to a depth of 2 inches
(5 cm), then level the soil surface with a
bed preparation rake to achieve a smooth
seedbed.

Using a Seeder

**Use a Six-Row Seeder and make
sure seeds are buried only lightly,
½ to ¾ inch (1–2 cm) deep.
Remember to sow successions for
a continuous harvest, seeding
every 10 to 15 days or so.**

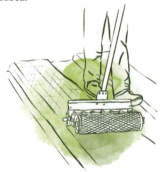

**Water with a gentle spray and
immediately set up a row cover
or insect netting on the crop;
this promotes germination and
will protect young leaves from
insects.**

Broadcast Seeding

1 For home gardeners sowing only small quantities, a seeder may not be necessary. You can broadcast seed radishes or even use the fold of the seed packet to pour them into a shallow furrow.

2 Lightly rake the soil, working very superficially to avoid burying the seeds too deep, then tamp it with the back of the same tool or with a tamper. Finish by watering the bed with a gentle spray.

Tip from Jean-Martin Fortier

As soon as the seeds have germinated and the first leaves have developed, you must water the crop regularly, never allowing the soil to dry out. Otherwise the radishes may become pungent. If the weather is hot and dry, consider providing shade by using, for example, insect netting directly over the foliage or supported by metal hoops.

Maintenance

Weeding

Radishes do not tolerate weed pressure. If there are any weeds, remove them by hand, going through the crop often to pull them out before they harm the radishes. When this crop is sown densely, its soil coverage will allow few weeds to germinate. For bigger crops and to loosen the soil, you can run a flex tine weeder down the bed as soon as the seedlings have 3 or 4 leaves, making sure not to pull out any radish shoots.

Plant Protection

Radishes are vulnerable to various cabbage flies and flea beetles. One way to act preventively is to establish a crop rotation.

Tip from Jean-Martin Fortier

Watch out for birds and slugs as they are fond of tender radish shoots, especially a few days after germination. You can set up insect netting over hoops to keep the birds away from the crop. To deter gastropods, lay a few wooden planks on the bed to create sheltered spaces where slugs and snails will take refuge, drawn in by the cool shade. Then simply get rid of them.

Harvest

Spring radishes are harvested 15 to 30 days after seeding, depending on the season and the variety. Harvest on demand and according to your needs when roots are the size of a hazelnut (round radishes) and when they are 3 to 4 inches (8–10 cm) long (oblong radishes). You can opt to pull out the entire bed, freeing up space for another crop, or harvest only the best radishes, providing more room for smaller ones to grow. However, with this latter approach, the bed will be occupied for longer, to the detriment of the following crop.

Lifting the Crop

Harvest after rain or watering so that the soil will be loose and the roots will pull out without any damage. To lift a radish, grasp the base of the stems with one hand, then twist and pull the root from the soil. Gather the radishes into bunches with a rubber band, then rinse off the soil.

Tip from Jean-Martin Fortier

Spring radishes should not be kept but must be eaten right away because they quickly become spongy. Radishes are a cool-weather vegetable, so if a dry summer is forecasted or your region has a very hot climate, it's best not to seed summer crops as the roots may become pungent and the plants will quickly go to seed.

Winter Radishes

———————●———————

"Winter radishes are easy to grow and keep well all winter long. They are one of our most popular winter crops, especially daikon radishes."
Jean-Martin Fortier

Planting

Seeding Under Shelter

Winter radishes are a fall crop because growing them at this time of year is more in keeping with ideal growing conditions (cooler temperatures) and limits the likelihood of bolting. It is best to sow them under cover and then transplant seedlings into the field around midsummer. Crops can then be harvested 50 to 60 days later, in the fall.

1 Sow 1 seed per cell in plug flats about 30 days before your intended transplant date. Seed at a very shallow depth, about ¼ inch (5 mm).

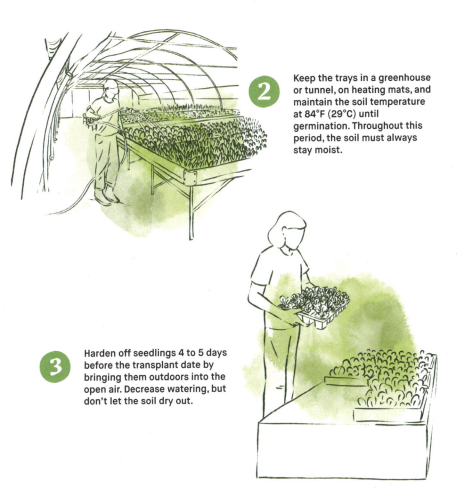

2 Keep the trays in a greenhouse or tunnel, on heating mats, and maintain the soil temperature at 84°F (29°C) until germination. Throughout this period, the soil must always stay moist.

3 Harden off seedlings 4 to 5 days before the transplant date by bringing them outdoors into the open air. Decrease watering, but don't let the soil dry out.

Tip from Jean-Martin Fortier

To save time and avoid the very tedious thinning process after direct seeding in the field, we prefer to sow winter radishes into plug flats or Paperpots.

Preparing the Soil

If there was a summer crop in the bed, remove the silage tarp (occultation tarp) and use a rake to pull out debris and clean up the surface. Loosen the soil with a broadfork, then spread chicken manure and compost. Next, use a power harrow set to a depth of 2 inches (5 cm) and level the soil with a bed preparation rake.

Tip from Jean-Martin Fortier

Winter radishes spend more time in the ground than spring radishes, so they need more fertilization. Use compost and pelleted chicken manure. This fertilization provides a nitrogen input that is quickly absorbed and necessary to establish the crop. With winter radishes, the fertilizer added before transplanting is the only input the crop will receive.

Planting

1 Mark your rows and spacings by running a row marker roller down each prepared bed. Plan for 3 rows per bed, set 8 to 10 inches (20–25 cm) apart, with an in-row spacing of about 8 inches (20 cm).

2 Drop the plants where the row marker roller lines intersect. One person should focus on this task, while another is responsible for planting, both occurring at the same time.

3 With a dibber or hand trowel, make a hole and bury the root ball, then tamp the soil with your hands. Water the crop thoroughly, directing a gentle spray around the plant to help the soil settle.

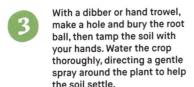

Care After Planting

After you finish transplanting the radish, set up metal hoops on both sides of the beds and immediately lay insect netting over the entire crop. Make sure the netting provides a perfect seal so that flies and flea beetles can't reach the foliage.

Maintenance

After planting is complete, you have a few more operations to carry out, and then the crop becomes fairly hands-off, requiring little work aside from monitoring irrigation and pest damage.

Irrigation

To avoid manual watering, the wise choice is to set up a drip irrigation system or sprinklers as a regular supply is required. Many winter radish roots do not grow deep, so even the slightest lack of water or dry spell can cause cracking. Lack of water also makes them pungent, affects the texture of the harvest, and causes root malformations.

Weeding

Start weeding the beds around 10 days after planting to eliminate emerging weeds. Then, if needed, repeat the operation 10 to 12 days later to suppress any regrowth. This should ideally be done with a stirrup hoe between the rows and around each plant. Eventually winter radishes develop foliage that provides enough ground cover to prevent weed growth.

Harvest

Winter radishes are ready to be harvested as early as 50 days after transplanting, depending on the size you want: about 6 to 8 inches (15–20 cm) for long winter radishes, and about the size of a ping-pong ball or tennis ball for round radishes. Either way, cautious growers should pull them out before the first fall frost.

 1 Clear away the foliage and locate a round root so you can grasp all the stems of a single plant. Twist the radish while pulling to remove it from the soil. Use a broadfork to lift the long roots.

2 If the radishes will be eaten right away, remove the foliage while keeping 1.5 to 2 inches (4–5 cm) of the stems, and wash the root. For longer storage, remove all the foliage and keep only ½ inch (1 cm) of the stem.

If you plan to store radishes, put them into crates, unwashed.

Tip from Jean-Martin Fortier

In mild climates, you can leave winter radishes in the field and harvest them on demand. This approach requires the roots and foliage to be protected from frosts. Lay a row cover over the entire crop or consider growing it under a low tunnel, then harvest radishes until the end of December.

Storage

After pulling winter radishes, store them unwashed and in crates, in a ventilated dark space kept at roughly 36°F to 40°F (2–4°C). A cold room is ideal as it will allow roots to keep for up to 4 months.

Home gardeners can use a cellar that provides the same conditions. If you don't have a cellar, a root clamp can deliver good results.

1 Brush the soil off the radishes, then put them on a layer of sand at the bottom of a crate.

2 Spread more sand over them, then add another layer of radishes, lined up side by side. And so on... Place your crates near a north-facing wall, and bury them in well-drained light soil or wrap them in straw and leaves covered with garden felt or burlap.

Focus on Daikon Radishes

Daikon radishes, also called Korean radishes and Japanese radishes, can be white or colored, and have a mild and slightly sweet taste. Long, semi-long, and round varieties like Alpine (white), KN-Bravo (purple), Red Meat (melon radish), and Luobo Vert are quite popular in Japan where this root vegetable is as important as the potato is in North America and the UK. In Europe, these radishes are increasingly popular, both at markets catering to individual consumers and in the restaurants of Michelin Star chefs.

While they are eaten raw, fermented, and cooked in Japan and Korea, elsewhere they are primarily used raw, in salads, and as a crudités.

There is such a diversity in radish varieties that they can be grown almost all year-round: in tunnels or under cold frames in winter and spring for a fresh harvest, and in the field from summer to fall for storage radishes. Daikon can be cultivated like winter radishes, with crop successions sown into plug flats kept under cover to be transplanted into the field, or direct sown into open ground. The soil must be deep and loose so that the plants can develop regular, uniform roots.

Data Sheet

COMMON NAMES: Black salsify, scorzonera, black oyster plant, Spanish salsify, viper's grass.
SCIENTIFIC NAME: *Scorzonera hispanica.*
FAMILY: Asteraceae.
REQUIREMENTS: Scorzonera is very cold hardy and undemanding. It grows well in deep loose soils that are free of rocks, so the taproot can grow unhindered. The crop is best suited to rich, warm soils with good sun exposure.

SPACING: 3 rows per bed, with in-row spacing of 2.5 to 3 inches (6–8 cm).
SEEDING: From March to April.
DAYS TO MATURITY: 7 to 8 months.
ENEMIES: While scorzonera is hardy and not susceptible to diseases and parasites, it is a target for rodents, which are fond of its roots in late summer.
VARIETIES: Pilotis, Géante Noire de Russie (aka Giant Black Russian, which has edible young leaves) and Hoffmanns (can be eaten raw).

NOTE FROM JEAN-MARTIN FORTIER
On farm stalls, scorzonera and salsify are often mistaken for each other. It's easy to tell them apart by their color: scorzonera skin is black, while salsify skin is beige!

Scorzonera

Scorzonera grows wild in Southern European pastures. While it was identified in antiquity, cultivation only began in the 17th century, first in Spain and then throughout Europe. Today, it is increasingly popular, and primarily grown in Belgium, for canning.

Although it is a perennial and very cold hardy, scorzonera is grown as an annual vegetable over 1 year or as a biennial spanning 2 years. Its elongated green leaves grow from a long cylindrical root with black skin and white flesh, which releases a milky substance (latex) at the slightest cracking. Scorzonera is less fibrous than salsify and has a milder, more delicate flavor, which is why it has dethroned salsify on market stalls and in kitchens.

Rich in fiber, various vitamins, and potassium, cooked scorzonera presents flavors reminiscent of artichoke, asparagus, and parsnip. The root is peeled, then boiled, steamed, or sautéed and eaten in gratins, béchamel sauces, and soups. Sometimes, young greens and the root are eaten raw in salads.

Scorzonera

———————•———————

"Because scorzonera spends a long time in the ground, it's best to dedicate only a corner of your vegetable garden or a fraction of a bed to this crop, to avoid monopolizing too much space."

Jean-Martin Fortier

Planting

Seeding

Use a broadfork to loosen the soil at depth then level the surface. Seed 3 rows by hand or with a Jang seeder. Keep the soil moist while the crop germinates and until it becomes established. When seedlings have 2 leaves, thin the crop, leaving plants 3 inches (7 cm) apart. While the crop is growing, weed, cultivate, and water regularly during dry weather, but allow soil to dry out between waterings and don't overwater.

Harvest and Storage

When lifting the crop, in October, be careful because the roots are brittle. Your safest way is to first loosen and lift the soil with a spading fork or broadfork. Gently extract the roots by pulling on the foliage, which will be cut off before the crop goes into storage. Keep roots in a dark cellar in a mixture of sand and soil as scorzonera dries out quickly, losing some of its flavor. When stored this way, it will keep until spring.

Focus on Salsify (Vegetable Oyster)

Salsify *(Tragopogon porrifolius)*, belonging to the same family as scorzonera, is a wild perennial plant commonly found in the stony soils of the Mediterranean basin.

Although first identified in antiquity, it was not widely cultivated until the 16th century, especially grown in Italy and a little in France. Still, salsify never truly found its audience and was soon displaced by scorzonera. The root has a flavor resembling parsnips, after it is cooked in water and then fried or roasted in a gratin. Young salsify leaves can be eaten in salads or mixed into mesclun.

Seed salsify from March to May, in rows set 8 to 10 inches (20–25 cm) apart in loose, rich, cool soil. When seedlings have 3 or 4 leaves, thin the rows, leaving 3 to 4 inches (8–10 cm) between plants. For salsify, watering, weeding, cultivation, and mulching give rhythm to the spring and summer months as the crop is slow growing and takes up a lot of space. In late summer and fall, after they have been growing for more than 6 to 7 months, carefully pull the roots from the soil. You may do so on demand (for immediate consumption) or all at once (for storage in a cellar). Although salsify is cold hardy, it should only stay in the field if the bed is covered by a thick layer of straw or dead leaves.

Data Sheet

COMMON NAMES: Jerusalem artichoke, sunchoke, giralsole, sunroot, topinambur.

SCIENTIFIC NAME: *Helianthus tuberosus.*

FAMILY: Asteraceae.

REQUIREMENTS: Cold hardy and undemanding, Jerusalem artichokes grow in decent garden soil that is loose, not overly fertile, and well drained. They thrive in full sun and can tolerate periods of summer drought, which impact foliage but not tubers.

SPACING: 1 to 2 rows per bed set 14 to 16 inches (35-40 cm) apart, with an in-row spacing of 24 to 28 inches (60-70 cm).

PLANTING: March to April.

DAYS TO MATURITY: 180 to 200 days (about 6 months).

ENEMIES: Powdery mildew on foliage, and rodents, which are fond of underground organs left in the field over winter.

VARIETIES: Fuseau, Violet de Rennes (purplish-pink skin), and Patate

(larger tubers are easier to peel). Note: in Canadian French, patate is the word for "potato," so make sure you are in fact purchasing Jerusalem artichoke and not potato seed.

NOTE FROM JEAN-MARTIN FORTIER

Grow Jerusalem artichokes apart from other garden vegetables as their vigorous, even invasive, growth can stress nearby traditional crops. You should also plan to stake flower stalks as late-season wind and rain will knock them down to the ground.

Jerusalem Artichokes

A close cousin of *Helianthus strumosus*, Jerusalem artichokes are a root vegetable native to North America. It crossed the Atlantic in the mid-15th century, where it met great success with European nobility. Although Jerusalem artichokes were dethroned by potatoes in the 19th century, they reemerged in World War II—as a substitute for potatoes, which were in short supply—and was later forgotten yet again. Today, Michelin Star chefs value Jerusalem artichokes for their delicate and sweet flavors, a blend of artichoke and water chestnut, which are drawn out through boiling or steaming. Jerusalem artichokes are rich in fiber, vitamins (like vitamin B), iron, and potassium. The root contains inulin, a fermentable fiber that is beneficial for digestion (but beware—it can cause intestinal gas!).

The organ we eat is a tuber that forms on a rhizome and produces a mass of flowering stems that can grow to over 10 feet (2.5 m) tall. The plant's yellow flowers are reminiscent of miniature sunflowers, and its green leaves are soft and smooth to the touch, which differentiates them from *Helianthus strumosus*. Jerusalem artichokes can be pan-fried, baked in gratins, or cooked into a purée or soup.

Jerusalem Artichokes

"Long banished from household kitchens and market gardens, Jerusalem artichokes are quietly making a comeback thanks to Michelin Star chefs. It is an excellent vegetable for winter market gardening."
Jean-Martin Fortier

Planting

Preparing the Soil
Use a broadfork to loosen the bed, then apply a layer of compost and pelleted chicken manure (5-3-2). Rake the surface to remove coarse aggregates and level the soil.

Planting
Purchase seed tubers or use any saved from the previous harvest and plant them in holes dug every 20 inches (50 cm) for a single row, or every 28 inches (70 cm) and staggered for 2 rows in a bed. Plant each tuber 4 inches (10 cm) deep, fill in the hole, and water thoroughly.

When shoots are 8 inches (20 cm) tall, hill the plants to better anchor them and improve the stability of the stems.

Maintenance

Cultivate the soil for the first few weeks
after planting, then cover the bed with a
straw mulch to keep the ground cool and
allow for less frequent watering. If needed,
stake the stems.

Harvest

Lift the crop when the first frosts blacken
the leaves. Begin by cutting the stems down
to 16 inches (40 cm) and loosen the soil—if
it is not frozen—using a broadfork or
spading fork to harvest tubers on demand,
as long as the winter is not too harsh. As a
precaution, you may choose to pull the
entire crop, harvesting it all at once. Do this
in dry and sunny weather, so you can let the
harvest air-dry in the open before storing it.

Storage

After the tubers have dried (brush but do
not wash), place them in crates. Mix in sand
or soil and keep it slightly moist as they can
easily dry out. Store the crates in a dark,
ventilated, and frost-free cellar.
Alternatively, place tubers in a tightly sealed
plastic bag, kept in a cellar, monitoring for
any signs of mold (which would require
removing damaged tubers and repacking
the remainder).

Acknowledgments from Jean-Martin Fortier

I wish to thank the entire team at the Market Gardener Institute for encouraging me to pursue my mission, every day. A big thank-you also goes out to the Growers & Co. team, who pushes me to come up with new types of equipment! I especially want to acknowledge my partner, Maude-Hélène Desroches, who is an exceptional market gardener and a dear friend!

Acknowledgments from New Society Publishers

We extend a great thanks to Delachaux et Niestlé, the French publisher, for working with us to publish this English edition. Further thanks to the New Society Publishers team for producing the book and especially to Laurie Bennett for her meticulous attention to technical details and high-quality translation into English.

Acknowledgments from Delachaux et Niestlé

A big thank-you to Jean-Martin Fortier and his team at the Market Gardener Institute for this wonderful collaboration.

Our heartfelt thanks go out to Pierre Nessmann for putting us in touch with Jean-Martin, for thoroughly editing this collection, and for being so generous with his time. For this book, we owe him so much. We also wish to thank Flore Avram, whose illustrations give this collection a beautiful, simple character; to Grégory Bricout for graphic design that cleverly reflects Jean-Martin Fortier's spirit; and to Sandrine Harbonnier and Sabine Kuentz for their work on the text.

Reference Books

The Market Gardener Masterclass, www.themarketgardener.com
The Market Gardener: A Successful Grower's Handbook for Small-Scale Organic Farming, New Society Publishers, 2014.
Winter Market Gardening: A Successful Grower's Handbook for Year-Round Harvests, New Society Publishers, 2023.
Microfarms: Organic Market Gardening on a Human Scale, New Society Publishers, 2024.

Grower's Guides from the Market Gardener

Tomatoes: A Grower's Guide
Vegetable Garden Tools: A Grower's Guide
Root Vegetables: A Grower's Guide
Living Soil: A Grower's Guide

Coming Soon

Fruiting Vegetables: A Grower's Guide
The Well-Planned Vegetable Garden
Fall and Winter Vegetables
Produce Your Own Plants
Salads and Leafy Greens
Herbs
Perennial Vegetables

Translator: **Laurie Bennett**

About New Society Publishers

New Society Publishers is an activist, solutions-oriented publisher focused on publishing books to build a more just and sustainable future. Our books offer tips, tools, and insights from leading experts in a wide range of areas.

We're proud to hold to the highest environmental and social standards of any publisher in North America. When you buy New Society books, you are part of the solution!

At New Society Publishers, we care deeply about *what* we publish — but also about *how* we do business.

* This book is printed on **100% post-consumer recycled paper**, processed chlorine-free, with low-VOC vegetable-based inks (since 2002)
* Our corporate structure is an innovative employee shareholder agreement, so we're one-third employee-owned (since 2015)
* We've created a Statement of Ethics (2021). The intent of this Statement is to act as a framework to guide our actions and facilitate feedback for continuous improvement of our work
* We're carbon-neutral (since 2006)
* We're certified as a B Corporation (since 2016)
* We're Signatories to the UN's Sustainable Development Goals (SDG) Publishers Compact (2020–2030, the Decade of Action)

To download our full catalog, sign up for our quarterly newsletter, and to learn more about New Society Publishers, please visit newsociety.com.

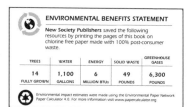

ENVIRONMENTAL BENEFITS STATEMENT

New Society Publishers saved the following resources by printing the pages of this book on chlorine free paper made with 100% post-consumer waste.

TREES	WATER	ENERGY	SOLID WASTE	GREENHOUSE GASES
14	1,100	6	49	6,300
FULLY GROWN	GALLONS	MILLION BTUs	POUNDS	POUNDS

Environmental impact estimates were made using the Environmental Paper Network Paper Calculator 4.0. For more information visit www.papercalculator.org

Certified **B** Corporation

SDG PUBLISHERS COMPACT

FSC www.fsc.org

MIX
Paper | Supporting responsible forestry
FSC® C016245

new society
PUBLISHERS
www.newsociety.com